T0239279

Series/Number 07/140

CONFIDENCE INTERVALS

MICHAEL SMITHSON
The Australian National University, Canberra

Quantitative Applications in the Social Sciences

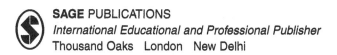
SAGE PUBLICATIONS
International Educational and Professional Publisher
Thousand Oaks   London   New Delhi

For information:

Sage Publications, Inc.
2455 Teller Road
Thousand Oaks, California 91320
E-mail: order@sagepub.com

Sage Publications Ltd.
6 Bonhill Street
London EC2A 4PU
United Kingdom

Sage Publications India Pvt. Ltd.
M-32 Market
Greater Kailash I
New Delhi 110 048 India

Library of Congress Cataloging-in-Publication Data

Smithson, Michael.
    Confidence intervals / Michael Smithson.
        p. cm. — (Quantitative applications in the social sciences; v. 140)
    Includes bibliographical references.
    ISBN 0-7619-2499-X (pbk.: alk. paper)
    1. Social sciences-Statistical methods. 2. Confidence intervals.
    3. Social sciences-Mathematics. I. Title. II. Sage university papers series.
    Quantitative applications in the social sciences; no. 140.
    HA31.2 .S59 2003
    519.5'38—dc21
                                        2002009707

06 07 08 09 10 11 12 13 14 10 9 8 7 6 5 4 3 2

Acquiring Editor:	C. Deborah Laughton
Editorial Assistant:	Veronica Novak
Production Editor:	Diana E. Axelsen
Copy Editor:	A. J. Sobczak
Typesetter:	C&M Digitals (P) Ltd.

When citing a university paper, please use the proper form. Remember to cite the Sage University Paper series title and include the paper number. One of the following formats can be adapted (depending on the style manual used):

SMITHSON, MICHAEL. (2003). *Confidence Intervals.* Sage University Papers Series on Quantitative Applications in the Social Sciences, 07-140. Thousand Oaks, CA: Sage.

OR

SMITHSON, MICHAEL. (2003). *Confidence Intervals.* (Sage University Papers Series on Quantitative Applications in the Social Sciences, series no. 07-140. Thousand Oaks, CA: Sage.

CONTENTS

SERIES EDITOR'S INTRODUCTION

A principal task of researchers is parameter estimation. From experimental or nonexperimental observations, researchers seek to ascertain the population value of a mean, a percentage, a correlation, a slope, or some other statistic. Commonly, a motivating question is whether the estimate is statistically significant. The answer "yes" encourages investigators to believe they are on to something, whereas the answer "no" makes them believe they are on the wrong track. Important as the significance test can be, however, it has been overvalued. In conventional application, as a tool to reject the null hypothesis, it merely informs us that the real number is probably not zero. It says nothing about how far it is from zero or, more generally, what range of values it might take. For such knowledge, which perhaps has more utility, one should turn to confidence intervals.

Suppose, for example, that Professor Jane Green, an occupational sociologist, estimates the correlation between annual income and job prestige, using responses from a random survey of the labor force in her community. The calculated correlation (Pearson's r) turns out to be .40. After application of a significance test (.05 level, two-tail), she rejects the null hypothesis and concludes that the population correlation is almost certainly not zero. But it comes as no surprise to her that the variables in fact seem related. Her real interest is the strength of the relationship. The estimate of .40 indicates at least moderate strength. However, taking into account sampling error, she understands that it might well be higher or lower than that. How much higher or lower?

According to a 95% confidence interval (CI), the range is [.2, .6]. Over repeated samples, construction of such a CI would capture the true value of the population correlation 95% of the time. She concludes, from this particular sample, that extreme values for the population correlation, say < .20 or > .60, are very unlikely. However, the plausible range of values from the CI suggests that the population correlation could run from a weak .20 to a rather strong .60. Professor Green realizes, then, that her estimate lacks precision. She would like a narrower band, in order to rule out the array of judgment calls from "weak" to "strong." She decides to repeat the study and to increase sample size.

In this book, Dr. Smithson observes that larger samples produce narrower CIs and therefore a narrower range of "plausible values" for the population parameter. For instance, in estimating a proportion, the CI is cut roughly in half if the sample size is quadrupled (assuming non-extreme proportions).

A confidence interval encompasses a significance test, because if the value of zero falls within the interval, the estimate is not statistically significant (at the level of 1–the CI level). However, a CI also gives information about the likely range of values, in the context of a particular sample size. As Dr. Smithson notes, in analyzing multiple studies concerning a certain effect, it allows the evaluator to do more than a simple vote-counting of how many studies found significance versus how many did not. Comparison of confidence intervals over the various studies may lead to different, and more accurate, conclusions concerning the presence and magnitude of an effect.

The monograph at hand is a tour de force, explaining how to construct and interpret confidence intervals for almost every conceivable applied statistic in ANOVA, regression, or categorical data analysis. Empirical examples are plentiful and clear. The CI calculated by most statistical packages assume standardizable distributions, such as the normal distribution. Generously, Dr. Smithson offers to provide readers with special routines that can be used for nonstandardizable distributions, such as the noncentral t. This encourages the application of CI, which is all to the good. Across the social science disciplines, confidence intervals have been explored much too little.

— *Michael S. Lewis-Beck*
Series Editor

CONFIDENCE INTERVALS

MICHAEL SMITHSON
The Australian National University

1. INTRODUCTION AND OVERVIEW

This monograph surveys methods for constructing confidence intervals, which estimate and represent statistical uncertainty or imprecision associated with estimates of population parameters from sample data. A typical example of a confidence interval statement is a pollster's claim that she or he is 95% confident that the true percentage vote for a political candidate lies somewhere between 38% and 44%, on the basis of a sample survey from the voting population. Pollsters often refer to the gap between 38% and 44% as the "margin of error." In statistical terms, the interval from 38% to 44% is a 95% *confidence interval*, and 95% is the *confidence level*. The pollster's claim actually means that she or he has a procedure for constructing an interval that, under repeated random sampling in identical conditions, would contain the true percentage of the vote 95% of the time. We will examine this more technical meaning in the next chapter.

This interval conveys a lot of information concisely. Not only does it tell us approximately how large the vote is, but it also enables anyone so disposed to evaluate the plausibility of various hypothetical percentages. If the previous election yielded a 39% vote for this candidate, for instance, then it is not beyond the bounds of plausibility (at the 95% confidence level) that the candidate's popularity has remained the same. This is because 39% is contained in the interval from 38% to 44% and therefore is a plausible value for the true percentage vote. That said, we also cannot rule out the possibilities of an increase by as much as 5% or a decline by as much as 1%.

The confidence interval also enables us to assess the capacity of the poll to resolve competing predictions or hypotheses about the candidate's popularity. We can rule out, for instance, a hypothesis that the true percentage is 50%, but we cannot rule out hypothetical values of the percentage vote that fall within the 38%-44% interval. If, for example, the candidate needs to gain a clear majority vote to take office, then this poll is able to rule that out as implausible if the election were held on the same day as the poll (assuming that a 95% confidence level is acceptable to all concerned). If, on the other hand, the candidate needs only a 4% increase to take office, then the confidence interval indicates that this is a plausible possibility. In fact, as we will see in Chapter 2, a confidence interval contains all the hypothetical values that cannot be ruled out (or rejected). Viewed in that sense, it is much more informative than the usual significance test.

This monograph refers to a fairly wide variety of statistical techniques, but many of these should be familiar to readers who have completed an undergraduate introductory statistics unit for social science students. Where less familiar techniques are covered, readers may skip those parts without sacrificing their understanding of the fundamental concepts. In fact, Chapters 2-4 and 7 cover most of the fundamentals. Chapter 2 introduces the basis of the confidence interval framework, beginning with the concepts of a sampling distribution and a limiting distribution. Criteria for "best" confidence intervals are discussed, along with the trade-off between confidence and precision (or decisiveness). The strengths and weaknesses of confidence intervals are presented, particularly in comparison with significance tests.

Chapter 3 covers "central" confidence intervals, for which the same standardized distribution may be used regardless of the hypothetical value of the population parameter. Many of these will be familiar to some readers because they are based on the t, normal, chi-square, and F distributions. This chapter also introduces the transformation principle, whereby a confidence interval for a parameter may be used to construct an interval for any monotonic transformation of that parameter. Finally, there is a brief discussion of the effect that sampling design has on variability and therefore on confidence intervals.

Chapter 4 introduces "noncentral" confidence intervals, based on distributions whose shape changes with the value of the parameter being estimated. Widely applicable examples are the noncentral t, F, and χ^2 (chi-squared) distributions. Confidence intervals for the noncentrality parameters associated with these distributions may be converted into confidence intervals for several popular effect-size measures such as multiple R^2 and Cohen's d.

Chapters 5 and 6 provide extended examples of the applications of confidence intervals. Chapter 5 covers a range of applications in ANOVA and linear regression, with examples from research in several disciplines. Chapter 6 deals with topics in categorical data analysis, starting with univariate and bivariate techniques and proceeding to multi-way frequency analysis and logistic regression.

Chapter 7 elucidates the relationship between the confidence interval and significance testing frameworks, particularly regarding power. The use of confidence intervals in designing studies is discussed, including the distinctions arising between considerations of confidence interval width and power. Chapter 8 provides some concluding remarks and brief mentions of several topics related to confidence intervals but not dealt with in this monograph, namely measurement error, complex sample designs, and meta-analysis.

I have received useful advice from many colleagues and students on drafts of this monograph. I am especially indebted to John Beale, Geoff Cumming, Chris Dracup, John Maindonald, Craig McGarty, Jeff Ward, and the students in the ACSPRI Summer School 2001 Confidence Interval Workshop for detailed and valuable ideas, data, criticism, and error detection. Of course, I am solely responsible for any remaining errors or flaws in this work.

2. CONFIDENCE STATEMENTS AND INTERVAL ESTIMATES

Let us return to the example confidence statement by the pollster, namely that she is 95% confident that the true percentage vote for a political candidate lies somewhere between 38% and 44%, on the basis of a sample survey from the voting population. Her requirements to make this statement are identical to those for estimating a population parameter with a sample statistic, namely a statistical model of how the sample statistic is expected to behave under random sampling error. In this example, the population parameter is the percentage of the voters who will vote for the candidate, but we could be estimating any statistic (e.g., a mean or the correlation between two variables).

Let us denote the population parameter by θ, whose value is unknown. We may define confidence intervals for values of θ given a *confidence level* of $100(1 - \alpha)\%$, where α lies between 0 and 1, and a sample size of N. Confidence intervals may have an upper limit or a lower limit, or both. A $100(1 - \alpha)\%$ *upper confidence limit* (U) is a value that, under repeated random samples of size N, may be expected to exceed θ's true value $100(1 - \alpha)\%$ of the time. A $100(1 - \alpha)\%$ *lower confidence limit* (L) is a value that, under repeated random samples of size N, may be expected to fall below θ's true value $100(1 - \alpha)\%$ of the time. The traditional two-sided confidence interval uses lower and upper limits that each contain θ's true value $100(1 - \alpha/2)\%$ of the time, so that together they contain θ's true value $100(1 - \alpha)\%$ of the time. The interval often is written as [L, U], and sometimes writers will express the interval and its confidence level by writing $\Pr(L < \theta < U) = 1 - \alpha$.

The limits L and U are derived from a *sample statistic* (often this statistic is the sample estimate of θ) and a *sampling distribution* that specifies the probability of getting each possible value that the sample statistic can take. This means that L and U also are sample statistics, and they will vary from one sample to another. To illustrate this derivation, we will turn to the pollster example and use the proportion of votes instead of the percentage. This conversion will enable us to use the normal distribution as the sampling distribution of the observed proportion, P. Following traditional notation that uses Roman letters for sample statistics and Greek letters for population parameters, we will denote the sample proportion by *P* and the population proportion by Π. It is customary for statistics textbooks to state that for a sufficiently large sample and for values of Π not too close to 0 or 1, the sampling distribution of a proportion may be adequately approximated by a normal distribution with a mean of Π and an approximate estimate of the standard deviation s_p, where

$$s_p = \sqrt{\frac{P(1 - P)}{N}} \qquad [2.1]$$

Another somewhat more general way of putting this is that the normal distribution is the *limiting distribution* of a sample proportion as *N* increases, due to the Central Limit Theorem.

This normal approximation is very convenient and before the advent of computers was a practical necessity as well. However, readers should bear in mind that large samples and values of *P* close to .5 are needed for the approximation to work well. These days, computers have rendered this approximation obsolete, but it is still popular and we will use it here. To construct a 100(1 − α)% confidence interval given a sample statistic *P*, we need only find the number of standard errors above and below *P* required to slice α/2 from the tails of the normal distribution. Denote this number by $z_{\alpha/2}$. The lower limit of the confidence interval is the mean of a normal distribution for which *P* slices α/2 from the upper tail, and the upper limit of the confidence interval is the mean of a normal distribution for which *P* slices α/2 from the lower tail. The resulting confidence interval for Π may be written as [*P* − *w*, *P* + *w*], where *w* is the *half-width* of the interval and is defined by

$$w = (z_{\alpha/2})(s_p). \qquad [2.2]$$

The pollster collected a random sample of 1,000 responses, of which 410 were a vote for the candidate concerned. So, $P = 410/1,000 = 0.41$ and

$$s_p = \sqrt{\frac{.41(1 - .41)}{1,000}} = 0.01555.$$

A 95% confidence level requires that $\alpha = .05$ and therefore that $\alpha/2 = .025$ be sliced from the lower and upper tails of the normal distribution. The required number of standard errors above and below P to do this is $z_{\alpha/2} = 1.96$. The half-width of the confidence interval is therefore

$$w = (z_{\alpha/2})(s_p) = (1.96)(0.01555) = 0.0305.$$

The lower limit is $L = P - w = .41 - .0305 = .38$, and the upper limit is $U = P + w = .41 + .0305 = .44$.

The 95% confidence interval for Π is therefore [.38, .44] or [38%, 44%]. This interval also may be written as $P \pm w$ because this interval is by definition symmetrical around P, and many textbooks adopt this convention (e.g., Howell, 1997; Lockhart, 1998). Moreover, the method of constructing confidence intervals provided here and in such textbooks implicitly assumes that symmetry by referring to a single normal distribution with two equal-sized tails sliced off, as shown in Figure 2.1. In fact, the one-distribution method assumes that the sampling distribution retains the same shape regardless of the value of the population parameter.

However, it is crucial to realize that confidence intervals are not always symmetrical around sample statistics and that their sampling distributions do not always have the same shape for different values, despite the fact that many textbooks convey this impression. Indeed, it is somewhat misleading to think of a confidence interval as being constructed "around P" even though it will always contain P's value. To illustrate this point, let us consider the exact sampling distribution for the proportion, namely the binomial distribution, and a method of constructing confidence intervals that does not rely on the assumptions of symmetry or constant shape.

Suppose we are observing 15 independent trials that have only two possible outcomes, A and ~A. If we know that the true probability of A, denoted here by Π, is .7, then the binomial distribution associated with $\Pi = .7$ tells us the probability of observing event A 0, 1, 2, ..., or 15 times out of the 15 trials. It is the sampling distribution for P when $\Pi = 0.7$ and $N = 15$.

If we do not know what Π is, we can nevertheless use such sampling distributions to construct a confidence interval for Π based on an observed

6

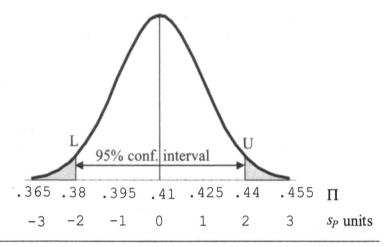

Figure 2.1   One-Distribution Representation

proportion of A's. Suppose we have observed 15 trials and 5 A's have occurred, so our observed proportion is $P = 5/15$. Now suppose we would like to construct a 90% confidence interval for Π. The way to do this is to find a lower limit on Π so that the probability of observing $P = 5/15$ or higher is .05 and an upper limit on Π so that the probability of observing $P = 5/15$ or lower is also .05.

Figure 2.2 displays the required sampling distributions in this example, one for $\Pi = .1417$ and the other for $\Pi = .5775$. We will return shortly to the question of where .1417 and .5775 came from. The heights of the bars in the sampling distribution for $\Pi = .1417$ from 5/15 on up are the probabilities of getting $P = 5/15$ or *higher* if Π actually is .1417. Those probabilities add up to .05. If we chose a hypothetical value of Π less than .1417, then we would be able to reject that hypothesis, given $P = 5/15$. According to our earlier definition, L = .1417 is a lower confidence limit on Π because under repeated random samples of size 15, we would expect an L chosen in this fashion to fall below Π's true value 95% of the time.

Likewise, the heights of the bars in the sampling distribution for $\Pi = .5775$ from 5/15 on down are the probabilities of getting a sample statistic $P = 5/15$ or *lower* if the population parameter Π actually is .5775. Those probabilities also add up to .05. If we hypothesized a value of population Π greater than .5775, then we would be able to reject that hypothesis, given a sample statistic $P = 5/15$. According to our earlier definition, U = .5775 is an upper confidence limit on Π because under repeated random samples of size 15, we would expect a U chosen in this fashion

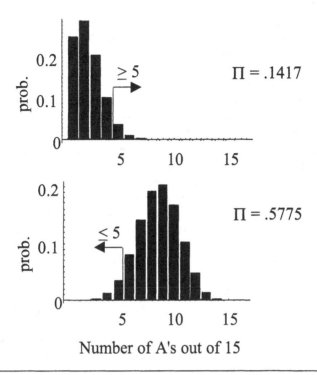

Figure 2.2   Two Binomial Distributions

to exceed Π's true value 95% of the time. So we have a 90% confidence interval for Π, namely [.1417, .5775].

Now, we cannot truthfully say that the probability of Π being somewhere between .1417 and .5775 is .9. However, according to the definition at the beginning of this chapter, we can say that if we repeatedly sampled 15 trials many times under identical conditions and followed the procedure outlined above, we could expect on average that our confidence intervals would contain the true value of Π at least 90% of the time. Imagine that Π really is .4, for instance, and we repeatedly draw samples of size 15. Each time we observe 2 or fewer A-events out of 15, our confidence interval will fail to include .4. It turns out that the 90% confidence interval based on $P = 2/15$ is [.0242, .3634] and thus excludes .4, whereas an interval based on $P = 3/15$ is [.0568, .4398] and so includes .4. But a standard binomial table tells us that if $\Pi = .4$, then the probability of observing 2 or fewer A-events out of 15 is less than .05 (about .0271, in fact).

By the same argument, every time we observe 10 or more A-events out of 15, our confidence interval will fail to include .4, because the 90%

confidence interval based on $P = 10/15$ is [.4225, .8583]. But the probability of observing 10 or more A-events out of 15 when $\Pi = .4$ is also less than .05 (about .0338). So, overall, we could expect our confidence intervals to include .4 more than 90% of the time.

How is this second method of constructing confidence intervals related to the first method that used the normal distribution? The first method actually uses a shortcut that is permissible (provided that the sample is large enough and the value of P sufficiently far from 0 or 1 for the normal approximation to be appropriate!) because the normal distribution is symmetrical and retains the same shape regardless of the value of Π. We may replicate the results there using the second method, thereby showing that the two methods are equivalent under symmetry and constant shape.

Figure 2.3 shows two normal distributions, one being the sampling distribution of P when Π is 1.96 standard errors below $P = .41$ and the other being the sampling distribution when Π is 1.96 standard errors above P. Because we know that $s_P = .01555$, we can find the corresponding lower and upper confidence limits on Π, which are $L = .41 - (1.96)(.01555) = .38$ and $U = .41 + (1.96)(.01555) = .44$, respectively. By definition, the probability of observing $P = .41$ or higher when the lower normal distribution holds is .025, and the probability of observing $P = .41$ or lower when the upper normal distribution holds also is .025. So the two-distribution method arrives at the same result as the one-distribution method. Unfortunately, the one-distribution method has limited generality and can be somewhat misleading. An earlier example of an introductory (though brief) treatment of confidence intervals that uses the two-distribution representation is Henkel's (1976) monograph on significance tests in the Sage Quantitative Applications in the Social Sciences (QASS) series (No. 4, p. 74).

It is worth contrasting this procedure with the "exact" interval we obtained using the binomial sampling model. In both cases, we could think of the procedure as sliding a sampling distribution up and down the axis by changing the value of Π until we find the appropriate lower and upper limits for the confidence interval. However, because the binomial distribution has a unique shape for every value of Π, we need a computer program (or, in the past, very patient humans) to iteratively home in on these limits. The reason we do not need this computationally intensive procedure for a confidence interval based on the normal approximation is that the normal distribution does not change shape with its location, and so we may solve for those limits. Prior to the wide availability of computing power, exact confidence intervals for the proportion for small samples were painstakingly hand-calculated and tabulated. Confidence intervals whose underlying sampling distributions are not symmetrical and not a constant shape were labor-intensive and impractical without computers, so they were (and still are)

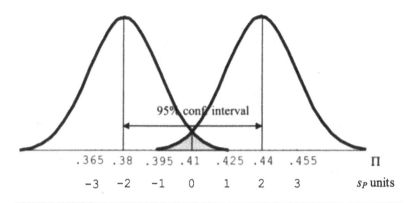

95% conf interval

| .365 | .38 | .395 | .41 | .425 | .44 | .455 | Π |

| -3 | -2 | -1 | 0 | 1 | 2 | 3 | s_P units |

Figure 2.3 Two-Distribution Representation

neglected in textbooks and, ironically, popular statistical computer packages. For the reader interested in the origins of the confidence interval approach, the framework was introduced almost in its entirety by Neyman in the 1930s (Neyman, 1935, 1937).

Decisions and Criteria for Confidence Intervals

Alert readers will have noticed that several arbitrary choices have been made in the construction of both kinds of intervals. First, we have assumed that the interval is *two-sided*, that is, with a lower and an upper limit. *One-sided* confidence intervals may suit some research purposes, as we shall see later. For now, suppose an experimenter is interested only in whether a proportion exceeds a criterion value such as 1/2 and has observed $P = 12/15$. He or she might want a one-sided $100(1 - \alpha)\%$ confidence interval using a lower limit, in which case he or she would wish to find a lower value for Π for which the probability of observing P or higher is α. For example, a one-sided 95% confidence interval given $P = 12/15$ would have the lower limit .5602, whereas a two-sided 95% interval would have the limits [.5191, .9567]. The lower limit for a one-sided 95% confidence interval is just the lower limit for a two-sided 90% confidence interval, which is why the one-sided interval provides the greater lower limit.

Although we will not take a hard-and-fast position here on one-sided versus two-sided confidence intervals, three characteristics should inform a researcher's choice. One is the inability to state how precise our estimate is when using a one-sided interval (cf. Steiger & Fouladi, 1997). Because interval estimation often serves the purpose of making statements about precision, this often is a determining criterion. The second is the relationship between the one-tailed versus two-tailed traditional significance test arguments, one-sided versus two-sided power statements, and one-sided versus two-sided confidence intervals. The third is the more generous lower limit afforded by a one-sided interval. We will revisit these issues at several points in this monograph.

A second assumption regarding two-sided confidence intervals is that α be split evenly between the lower and upper limits. In practice, this is often a matter of convention, but in some circumstances, it is dictated by criteria that statisticians have used for determining the "best" possible confidence interval procedure. Briefly, these criteria include exactness, conservativeness, unbiasedness, uniform greatest accuracy, and narrowness.

A $100(1 - \alpha)\%$ confidence interval is *exact* if it can be expected to contain the relevant parameter's true value $100(1 - \alpha)\%$ of the time. Often, exact intervals are not available or are inconvenient to calculate, and approximate intervals are used instead. If the rate of coverage of an approximate interval is greater than $100(1 - \alpha)\%$, then the interval is *conservative*; if the rate is less, then the interval is *liberal*. Generally, conservative intervals are preferred over liberal ones.

Given two or more intervals with identical coverage rates, several other criteria may be used in some circumstances to determine which is best. The simplest criterion is *narrowness*—narrower intervals are more informative and therefore are preferred over wider ones whose coverage rate is identical. The $100(1 - \alpha)\%$ interval that has the smallest probability of containing values other than the true parameter value is termed the *uniformly most accurate*. An *unbiased* confidence interval is one whose probability of including any value other than the parameter's true value is less than or equal to $100(1 - \alpha)\%$.

As a brief illustration, consider the normal approximation of a confidence interval for the proportion. It is not difficult to demonstrate that for high confidence levels, the shortest $100(1 - \alpha)\%$ confidence interval is one that allocates $\alpha/2$ to the lower limit and to the upper limit. If we allocate $\alpha/2 + \gamma$ to the lower limit, for example, then we must distribute $\alpha/2 - \gamma$ to the upper limit. The lower limit will be shifted by a greater amount than the upper limit if $\alpha/2$ is small, because in that region the normal curve steepens as we move toward the mean. For example, a 95% confidence interval with .025 allocated to each limit has a width of $(z_{.025} + z_{.025})(s_p) = (1.96 + 1.96)(s_p) = 3.92(s_p)$. A 95% interval with .04 allocated to the lower limit and .01 to the upper limit has a width of $(z_{.04} + z_{.01})(s_p) = (1.75 + 2.33)(s_p) = 4.08(s_p)$, and one with .0499 allocated to the lower limit and .0001 to the upper has a width of $5.37(s_p)$.

The third crucial choice concerns the confidence level. The 95% and 99% confidence levels are the most popular, but these simply reflect the popularity of the .05 and .01 Type I error rates in the significance testing tradition. The choice of confidence level is important because, all else being equal, the more confidence required, the wider the confidence interval. For a constant sample size and non-extreme values, confidence intervals for the proportion approximately double in width as we move from 75% to 99% in confidence level. The trade-off is between confidence and informativeness or decisiveness. A 100% confidence interval for the proportion from 0 to 1 is trivial because it is utterly uninformative. On the other hand, a 50% confidence interval is clearly too risky for most decision-making purposes regardless of the decisiveness it provides. Informativeness, on the other hand, does not require high confidence levels. In the 1930s, psychological researchers used 50% intervals to describe the "probable" latitude of error in their estimates, and classic texts such as Kempthorne and Folks (1971) recommend that researchers examine several confidence levels to ascertain their effects on the interval widths.

The other main determinant of the width of a confidence interval is sample size. Naturally, larger sample sizes yield narrower intervals. Given a constant confidence level, confidence intervals for non-extreme propor- tions approximately halve in width if we quadruple the sample size. We will examine both issues of confidence level and sample size more closely later, especially in relation to questions of precision, power, and Type I error rates when designing studies. For now, it should be noted that most fields of study do not have well-established criteria for deciding on confidence levels or interval widths. In practice, most of these decisions are dominated by conventions.

Confidence levels for multiple confidence intervals sometimes must be considered carefully by the researcher. Just as the Type I error rate for mul- tiple significance tests on the same data exceeds the error rate for each individual test, the *family-wise* coverage rate for simultaneous confidence intervals falls short of that for each interval on its own. Given k confidence intervals with confidence levels $1 - \alpha_i$ ($i = 1, 2, \ldots, k$), the Bonferroni inequality stipulates that the family-wise confidence level for these k inter- vals may be as low as $1 - \sum \alpha_i$. In the simplest case where $\alpha_i = \alpha$ for all i, the lower bound on this confidence level is just $1 - k\alpha$.

If a family-wise confidence level of $1 - \alpha$ is desired, then the Bonferroni correction is simply to use a confidence level of $1 - \alpha/k$ for each interval. This is not the only way to achieve the desired family-wise confidence level. For many statistics, there are methods of constructing *confidence regions*, which are multivariate versions of simultaneous confidence inter- vals. Confidence regions and the Bonferroni correction both lead to wider intervals than one-at-a-time intervals, but one method may produce wider intervals than the other. In fact, the Bonferroni method becomes unacceptably conservative when k is large. A researcher wishing to establish a family-wise confidence level should therefore choose the method best suited for his or her goals. Simultaneous confidence intervals (or confidence regions) will not be dealt with further in this monograph, but the interested reader should consult Tukey (1991).

Why Confidence Intervals?

Confidence Intervals Versus Significance Tests

On all counts, the confidence interval seems clearly superior to the traditional significance testing approach. Significance tests focus on just one null hypothesis value, whereas confidence intervals display the entire range of hypothetical values of a parameter that cannot be rejected. Although the confidence interval can be used as a significance test, there is no need for it to be used in that way, and in fact such an interpretation willfully ignores most of the information being provided by a confidence interval.

This is not to ignore or minimize the relationship between confidence intervals and tests of significance. After all, the usual description of a confidence interval involves the "inversion" of a significance test. As Cox and Hinkley (1974, p. 214) succinctly put it: "To obtain 'good' $1 - \alpha$ upper confidence limits, take all those parameter values not 'rejected' at level α by a 'good' significance test against lower alternatives." The notion of "goodness" here can have technical meanings. For instance, Lehman (1986, pp. 90-91) shows that inverting a uniformly most powerful significance test yields a uniformly most accurate confidence interval.

Confidence intervals enhance comparisons between research replications and enable researchers to move toward cumulative knowledge based on replications. Significance tests are not readily able to do this and, as has been argued by various authors (e.g., Schmidt, 1996, and authors in Harlow, Mulaik, and Steiger, 1997), significance tests may actually blind researchers to cumulative evidence. Suppose we observe 40 trials in which either A or ~A can occur, and we are interested in whether A is the more likely of the two outcomes (a similar scenario is presented in more detail in Smithson, 2000). If we find there are 24 A's, our observed proportion is .6, and we are unable to reject the null hypothesis that the true proportion is .5 (using a Type I error rate of $\alpha = 0.05$). This evidence does not enable us to conclude whether A is the more probable outcome or not. Now suppose we replicate this study and observe 40 more trials under identical conditions, this time obtaining 27 A's. In this study, the observed proportion is .675 and we are able to reject the null hypothesis.

If we relied solely on significance-testing for these studies, we could be misled into thinking that the two studies constitute one "vote" for and one

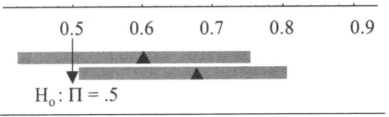

Figure 2.4   95% Confidence Intervals From Two Studies

against the null hypothesis. Confidence intervals give quite a different picture. An exact 95% confidence interval for the proportion from the first study turns out to be [.433, .751], and the 95% interval from the second study is [.509, .814]. As Figure 2.4 shows, these intervals overlap considerably, and the combined evidence indicates that the true proportion of A's is greater than .5.

Now imagine that we have run 25 more replications. In 15 of them, we cannot reject the null hypothesis, and in 10 we can. If we simply counted "votes" using significance tests, we could be forgiven for finally giving up on the hypothesis and believing that the evidence is equivocal at best. However, a meta-analytic approach would not lead us to that conclusion, nor would confidence intervals. As Figure 2.5 shows, almost all of our replications have found essentially the same effect, and the cumulative evidence for an effect is strong because the very large majority of confidence intervals are to the right of 0.5.

Actually, these "replications" are random samples of 40 from a population in which the probability of A occurring is .65. The confidence intervals in Figure 2.5 demonstrate that the reason for not being able to reject (correctly!) the null hypothesis very often is low power. Confidence intervals are very effective for this kind of exploration.

Significance tests can seduce researchers into ignoring statistical power, whereas confidence intervals make power salient by displaying the width (or imprecision) of the interval estimates. As several authors from Cohen (1962) onward have pointed out, in some subdisciplines, the majority of studies have low statistical power to detect effects of the size that researchers expect. Confidence interval techniques alert researchers to low power by producing wide intervals.

We have already seen how significance tests fail to contribute to cumulative knowledge, whereas confidence intervals pave the way. There are several reasons for this, but the main reason is that merely reporting a *p* value (or worse still, whether or not a result is significant at the .05 level) obscures all the factors that might determine significance, such as sample

14

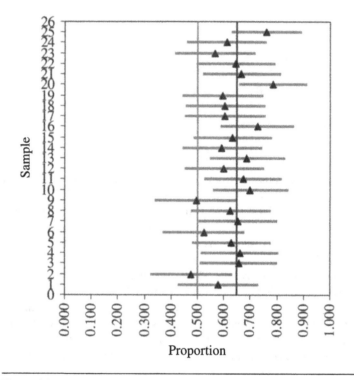

Figure 2.5 Confidence Intervals for 25 Replications

size and effect size. The confidence interval alerts the researcher to both of these, as well as enabling realistic comparisons between studies.

Consider a researcher who finds that the confidence interval in one study is much wider than, and includes, the confidence interval from another study. She or he is in a much better position than a researcher who knows only that the first study's result was nonsignificant and the second one significant. She or he immediately knows that both results are not clearly different and that she or he should look at sample sizes or the variances to account for the difference in precision between the two studies. As expressed by Wilkinson and the APA Task Force on Statistical Inference (1999), "Comparing confidence intervals from a current study to intervals from previous, related studies helps focus attention on stability across studies. . . . Collecting intervals across studies also helps in constructing plausible regions for population parameters" (p. 599).

Large samples pose a quandary for researchers using significance tests because even tiny deviations away from the null hypothesis will be found "significant." This is especially vexing when researchers are using "badness-of-fit" indices such as the chi-square-based measures in log-linear analysis or structural equations modeling. Almost paradoxically, a large sample will induce the researcher to reject a model when in fact it has a very close fit to the data. Confidence intervals do not have this problem. Large samples simply yield narrow confidence intervals and therefore narrow bounds on plausible values for a parameter or on plausible models for a data-set.

Finally, although significance tests are designed only for hypothesis testing, confidence intervals are useful for both hypothesis testing and exploratory research. In exploratory research, confidence intervals provide information about the (im)precision of sample estimates. They also assist in the generation of hypotheses by displaying the range of plausible values for a parameter, which in turn informs the researcher about the relative plausibility of competing predictions regarding that parameter.

Different disciplines have varied considerably in the extent to which confidence intervals are routinely reported in published research. Some of the social sciences have a long-standing tradition of using them, usually where the emphasis is on parameter estimation rather than hypothesis testing and often in the context of sample surveys. More experimentally oriented disciplines, on the other hand, traditionally have relied on significance tests and often eschewed confidence intervals altogether. This was the situation in medicine until about 10-15 years ago and still tends to be the case for psychology.

The advantages that confidence intervals have over null-hypothesis significance testing have been presented on many occasions to researchers in psychology. Early exponents of this viewpoint include Rozeboom (1960) and Meehl (1967), and more recent ones are Oakes (1986) and Schmidt (1996). Several recent commentators have wondered why psychological researchers have clung to significance testing for so long. However, standard practices in psychology do appear likely to change at last, given the conclusions of the APA Task Force on Statistical Inference (Wilkinson & APA Task Force on Statistical Inference, 1999). The guidelines make their position quite clear:

It is hard to imagine a situation in which a dichotomous accept–reject decision is better than reporting an actual *p* value or, better still, a confidence interval. . . . Interval estimates should be given for any effect sizes involving principal outcomes. Provide intervals for correlations and other coefficients of association or variation whenever possible (p. 599).

Similar arguments were produced in guidelines for medical researchers about 12 years earlier. Bailar and Mosteller (1988) advise researchers to present quantified findings with appropriate indicators of measurement error or uncertainty (such as confidence intervals). They admonish them to avoid sole reliance on statistical hypothesis testing and point out that confidence intervals offer a more informative way to deal with the significance test than a simple p value does. Confidence intervals for a mean or a proportion, for example, provide information about both level and variability.

Weaknesses and Limitations

Perhaps the most obvious difficulty with confidence intervals lies in how we interpret what the confidence statement means. Unfortunately, a 95% confidence interval for a proportion does not mean that the probability that the population proportion value lies inside that particular confidence interval is .95. The 95% confidence refers instead to the expected percentage of such intervals that would contain the population value, if we repeatedly took random samples of the same size from the same population under identical conditions.

This interpretive quandary is indeed a problem, if for no other reason than the considerable temptation to apply the confidence level to the specific interval at hand. As Savage (1962, p. 98) remarked under a somewhat different context, "The only use I know for a confidence interval is to have confidence in it." It is difficult to convey a confidence interval without at least implicitly leading the reader to apply the confidence level to the specific interval at hand. Cox and Hinkley (1974, p. 209) are inclined to regard this distinction as "legalistic." However, yielding to this temptation may result in incoherent inferences, as several authors have shown. A summary of the arguments involved and a survey of the relevant literature are available in Walley (1991, Section 7.5); the interested reader may follow his citations to the primary sources.

This difficulty is more consequential than it might seem at first, because the immediate use of a confidence interval for inductive empirical work (i.e., either for coming to a conclusion from a particular study or for making inferences based on a sample) does seem to require some measure of how certain we can be that the interval contains the true value of the parameter being estimated. There have been attempts to reinterpret confidence levels in ways that render them more useful for purposes of induction or inference. One approach (Cox & Hinkley, 1974; Kempthorne & Folks, 1971) interprets the confidence level as a measure of how "consonant" values contained in a confidence interval are with the sample statistic value.

Even if we do not fall prey to the temptation, according to critics from alternative schools of inference, the confidence interval framework shares some defects and limitations with its significance-testing cousin. These have been raised elsewhere (e.g., Kendall, 1949; Lindley, 1965), and their full explication lies beyond the scope of this work. Briefly, confidence interval techniques do not permit the user to take account of prior knowledge, which is certainly problematic for decision making or even cumulative knowledge. Bayesians are fond of pointing out how their approach does permit the consideration of prior knowledge and, moreover, enables confidence statements to be attached to specific intervals rather than long-run expectations.

However, one need not be Bayesian or even a trained statistician to find situations in which confidence intervals appear pathological or at least counterintuitive. Standard (even optimal) procedures can lead in some cases to confidence intervals that have no width, that cover the entire range of possible values, or whose limits are out of range. We will see situations of this kind in random-effects ANOVA models for two-sided confidence intervals for variance ratios, and other examples have been described that are relevant to practical statistical work (e.g., Cox & Hinkley's [1974] example of the ratio of two means).

A relatively minor but vexing practical problem is establishing meaningful guidelines for confidence levels and acceptable interval widths. This is not a problem for the confidence interval framework per se; the same could be said of establishing criteria for significance, power, effect size, posterior probability, and so on. The onus is on researchers to carefully consider the trade-offs between confidence, interval width, and sample size requirements in their research areas. They may in the end conclude that the samples available to them simply are incapable of affording them desirable interval widths and/or confidence levels (thereby resorting to meta-analysis), but at least this will be an informed outcome. Considerations along these lines are dealt with in Chapter 7.

The issues in this section have been raised to alert researchers to the fact that some aspects of the confidence interval framework are controversial and, in some instances, problematic for even practical purposes. After all, there are alternative schools of statistical inference that have ongoing disputes over fundamentals. Nevertheless, when all is said and done, confidence intervals do work well for a wide range of practical research purposes. Many confidence interval computational routines are simple and readily available, which accounts for their popularity. In some disciplines, confidence intervals are well-established as a standard part of research practice, and there are handbooks that inform researchers about how to use and interpret them (in medical research, for instance, see Altman, Machin, Bryant, & Gardner, 2000). For disciplines such as psychology, which are attempting to break the dominance of significance testing, the confidence interval approach is a natural alternative that does not demand a complete break with the older research literature. The fact that traditional significance testing can be taught in introductory statistics courses for researchers as a special application of confidence intervals (cf. Lockhart, 1998; Smithson, 2000) means that teachers do not have to disenfranchise students from the research literature or thesis supervisors who still adhere to significance tests. Whether the Neyman-Pearson confidence interval framework will be superseded by Bayesian or likelihood techniques is uncertain, but for now in several disciplines this approach appears relatively workable and robust (especially when augmented with bootstrapping and other resampling techniques).

3. CENTRAL CONFIDENCE INTERVALS

Central and Standardizable Versus Noncentral Distributions

Some of the most widely used confidence intervals are based on *central* distributions such as t, F, and χ^2, which would reflect reality if a null hypothesis were true. This is in good part because the desired confidence intervals can be obtained by a straightforward rearrangement of statistics based on these null hypothesis distributions. *Noncentral* distributions, on the other hand, are produced by effects of independent variables and/or deviations away from null hypothesis parameter values. One of the main practical differences between these two kinds of distribution is that for central distributions, for any specified sample size and degrees of freedom, the same standardized distribution may be used regardless of the hypothetical value of the population parameter. In fact, central distributions are a subclass of *standardizable* distributions that have this property. Noncentral distributions change shape depending on the value of the parameter, so a unique distribution must be computed for every parameter value under consideration.

The normal distribution is an example of a standardizable distribution. For any sample size and sample proportion value, the normal approximation of a confidence interval for the proportion requires only the standard normal distribution with a mean of 0 and standard deviation of 1. The binomial distribution, as we have seen, is not standardizable and changes shape with every value of the proportion. This property results in computationally intensive methods for constructing confidence intervals when using noncentral distributions.

This chapter deals with confidence intervals based on standardizable distributions (including central ones). These intervals will be called "central" confidence intervals, even though in some applications they are not restricted to central distributions in the strict sense. The first two sections cover confidence intervals for the mean and the difference between means, and for the variance and the ratio of two variances. The final section introduces methods for transforming statistics in ways that lend them to confidence interval estimation using standardizable distributions. Examples include the correlation coefficient and odds ratio.

Confidence Intervals Using the Central *t* and Normal Distributions

The confidence interval for the mean and the difference between means when the variance is unknown relies on the fact that the sampling distribution of the *t* statistic is a central *t* distribution. The *t* statistic here is defined as in Formula 3.1:

$$t_{df} = \frac{\bar{X} - \mu}{s_{err}}$$

[3.1]

where $df = N - 1$ (i.e., the degrees of freedom), $s_{err} = s/\sqrt{N}$ (i.e., the standard error of the mean), and μ is a hypothetical value for the population mean.

Formula 3.1 provides the wherewithal for a confidence interval for μ in the same manner as the normal approximation method for the proportion. First, we find the number of standard errors above and below the sample mean required to slice $\alpha/2$ from the tails of the t_{df} distribution. Denote this number by $t_{df;\alpha/2}$. The half-width of the interval is then defined as

$$w = (t_{df;\,\alpha/2})(s_{err}).$$

[3.2]

The resulting confidence interval for μ may be written as $[\bar{X} - w, \bar{X} + w]$. Where it is possible to avoid confusion, the *df* subscript may be dropped, so that we write $t_{df;\alpha/2}$ as $t_{\alpha/2}$.

Example 3.1: Confidence Interval for a Mean

If a random sample of 20 adults from a population whose mean IQ is 100 is given a treatment that is claimed to increase IQ, and their posttreatment mean IQ is 106.5 with a standard deviation of 15, what would a confidence interval tell us about the potential validity of the claim that the treatment is effective? The traditional significance test would assess whether the sample mean of 106.5 differs from the no-effect hypothesis mean of 100. If we used a criterion for Type I error of $\alpha = .05$, then our criterial $t_{\alpha/2} = 2.093$. The standard error is $s_{err} = s/\sqrt{N} = 15/\sqrt{20} = 3.354$, so from Formula 3.1 we have

$$t_{19} = (\bar{X} - \mu)/s_{err} = 6.5/3.354 = 1.938.$$

Because $t_{19} < t_{\alpha/2}$, we would be unable to reject the hypothesis that people exposed to this treatment have a mean IQ of 100.

A confidence interval, of course, tells us much more than this. From Formula 3.2, the half-width of the interval is $w = (t_{\alpha/2})(s_{err}) = (2.093)(3.354) = 7.020$. So the 95% confidence interval for μ is [106.5 − 7.020, 106.5 + 7.020] = [99.48, 113.52]. Not only is the hypothesis that $\mu = 100$ included in this interval, but so is any value of μ from 99.48 to 113.52. We can also see that the estimate is not very precise, because the confidence interval width is nearly as large as the sample standard deviation.

The normal approximation to a confidence interval for μ simply replaces $t_{\alpha/2}$ in Formula 3.2 for the half-width with $z_{\alpha/2}$, so that $w = (z_{\alpha/2})(s_{err})$. Strictly speaking, the normal distribution is appropriate only when the population standard deviation is known. However, for sufficiently large samples (at least 50), it is a reasonable approximation to the t distribution. Because the t distribution has fatter tails than the normal distribution, w is smaller for the normal approximation and therefore the confidence interval is narrower. In our example, we would use $z_{\alpha/2} = 1.96$ rather than $t_{\alpha/2} = 2.093$, so the half-width is $(1.96)(3.354) = 6.574$ instead of 7.020. This difference is too large to ignore for many practical purposes. For larger samples, the difference in half-widths decreases, however. If $N = 60$ instead of 20 in our example, then $t_{\alpha/2} = 2.001$, which is closer to $z_{\alpha/2} = 1.96$. The resulting half-width for the normal approximation is 3.795, quite close to the t-based half-width of 3.875.

Example 3.2: Repeated Measures Case

The setup for the one-sample case also applies to the repeated-measures (or dependent samples) situation, in which each case has two observations on the same variable. The mean and standard deviation are of the within-subject differences (i.e., the differences between the two observations for each case). Consider an example of the repeated-measures setup from Smithson (Smithson, 2000, pp. 194-196). The fictitious social facilitation study presented there involves measurements of 12 800-meter runners' times under two conditions (no one present vs. one person present) (see Table 3.1).

TABLE 3.1
800-Meter Times for Athletes

Athlete	Absent A	Present B	A – B
1	112.3	111.2	1.1
2	110.7	111.9	-1.2
3	106.1	105.3	0.8
4	115.3	112.9	2.4
5	109.8	107.4	2.4
6	108.9	109.1	-0.2
7	106.0	106.2	-0.2
8	107.4	106.3	1.1
9	114.3	111.2	3.1
10	111.1	109.6	1.5
11	109.5	108.2	1.3
12	112.2	112.5	-0.3
Mean	110.300	109.317	0.983
Standard deviation	2.966	2.641	1.284
Variance	8.800	6.976	1.649
Correlation			.902

Source: Adapted from Table 6.3 in Smithson (2000, p. 195).

The mean difference is 0.983 seconds faster for the someone-present condition, and the standard deviation of the differences is $s = 1.284$. So $s_{err} = s/\sqrt{N} = 1.284/\sqrt{12} = 0.371$. If we desire a 95% confidence interval for the mean difference, then for $df = N - 1 = 11$, $t_{\alpha/2} = 2.201$, and from Formula 3.2, we have $w = (t_{\alpha/2})(s_{err}) = (2.201)(0.371) = 0.817$. The 95% confidence interval is [0.983 − 0.817, 0.983 + 0.817] = [0.166, 1.800].

Again, the confidence interval provides a more informative picture than the one afforded by a t test. In this example, the test for no difference yields $t_{11} = 0.983/0.371 = 2.650$, which many researchers would report with its significance level $p = .0226$. They might even note that this is less than half the criterion $\alpha = .05$. But because the sample size is so small, the confidence interval tells us that if the effect exists, then our sample is reasonably probable, but the effect could plausibly be anything from 0.166 seconds to 1.8 seconds.

This is probably a good opportunity to examine a common intuition about confidence intervals for differences versus confidence intervals for the means of each measurement occasion. Some students (and perhaps researchers too!) seem to believe that in a situation such as this, where the mean difference between the someone-present and absent conditions appears to be significantly greater than 0, the confidence intervals of the means themselves should not overlap or should overlap only slightly. That is not generally the case, and in this example, the overlap between the two intervals is considerable.

The mean time in the absent condition is 110.300 seconds, and the standard deviation is 2.966. So $s_{err} = 2.966/\sqrt{12} = 0.856$ and, as before, $t_{\alpha/2} = 2.201$, so the 95% confidence interval for the mean in this condition is [108.415, 112.185]. The mean time in the someone-present condition is 109.317, and the standard deviation is 2.641, so $s_{err} = 0.762$ and the 95% confidence interval for the mean in this condition is [107.639, 110.995]. These intervals share a substantial majority of their respective ranges.

The difficulty is that if these intervals were tabulated or displayed graphically (as is frequently the case for repeated-measures data), they would convey a rather misleading impression about the differences between the two conditions. The correlation between A and B ($R = .902$) is behind this counterintuitive outcome. The variance of A − B is

$$s_A^2 + s_B^2 - 2Rs_A s_B$$

so the larger R is, the smaller this variance will be, and therefore the narrower the confidence interval for the mean difference.

Two Independent Samples

We now turn to the two independent samples setup, where the object is a confidence interval for the difference between two population means. Once again, this is linked to a t statistic, defined by

$$t_{df} = \frac{(\bar{X}_1 - \bar{X}_2) - (\mu_1 - \mu_2)}{s_{err}} \qquad [3.3]$$

where $df = N_1 + N_2 - 2$. Assuming that the variances for both populations are the same (homogeneity of variance), the standard error s_{err} is based on the pooled variance of the difference between two means. This pooled variance (s^2_{pooled}) is a weighted sum of the variances from each sample, where the weights are based on the size of the sample:

$$s^2_{pooled} = \frac{(N_1 - 1)s_1^2 + (N_2 - 1)s_2^2}{(N_1 - 1) + (N_2 - 1)}. \qquad [3.4]$$

The resulting pooled standard error is

$$s_{err} = \sqrt{s^2_{pooled}\left[\frac{1}{N_1} + \frac{1}{N_2}\right]}.$$ [3.5]

The confidence interval for $\mu_1 - \mu_2$ is thus $[\bar{X}_1 - \bar{X}_2 - w, \bar{X}_1 - \bar{X}_2 + w]$, where $w = (t_{\alpha/2})(s_{err})$ as in Formula 3.2.

In Chapter 5, we will apply the material from this section to ANOVA and regression. The primary applications in ANOVA are confidence intervals for contrasts. The most popular applications in regression concern confidence intervals for individual regression coefficients, which usually are provided in the regression output of standard statistical packages.

Example 3.3: Confidence Interval for Difference of Two Means

Windschitl and Wells (1998) report a study in which half of the 74 participants were told that "Randy cleans an average of 30 [classrooms] per week, Amy cleans 7, Laura cleans 5, Matt cleans 5, and Sylvia cleans 3," and the other half were told that "Randy cleans an average of 30 and Sylvia cleans an average of 20." All participants then evaluated the likelihood that Randy cleaned a particular classroom by choosing a response on a 21-point "verbal uncertainty" scale. The authors then claim that the mean likelihood rating for the 30-7-5-5-3 condition ($\bar{X}_1 = 13.9$, $s_1 = 2.5$) is higher than the mean for the 30-20 condition ($\bar{X}_1 = 11.9$, $s_2 = 1.5$) on the basis of the usual t test ($t_{72} = 4.00$, $p < .001$). They refer to this as a "robust" effect.

The difference between the sample means is 2.0, and we will construct a 95% confidence interval for this difference. Using their t value, from Formula 3.3 we get

$$s_{err} = (\bar{X}_1 - \bar{X}_2)/t_{72} = 2.0/4.00 = 0.5.$$

Because $t_{\alpha/2} = 2.289$, we have

$$w = (t_{\alpha/2})(s_{err}) = (2.289)(.5) = 1.145.$$

The confidence interval is therefore $[2.0 - 1.145, 2.0 + 1.145] = [0.855, 3.145]$. This amounts to a finding that the true difference between the mean ratings is somewhere between about 1 and 3 points on the 21-point scale.

Confidence Intervals Using the Central Chi-Square and F Distributions

The most popular confidence intervals using the central chi-square and F distributions concern variances and ratios of variances. These have many applications in ANOVA and related techniques, as do their noncentral counterparts. We begin with a confidence interval for the variance.

Confidence Intervals for Variances

Suppose we have a collection of K independent normally distributed random variables $\{Y_i, i = 1, \ldots, K\}$ whose means are all 0 and whose variances are all σ^2. Then the sum of squares of these variables divided by σ^2 (i.e., $\Sigma Y_i^2/\sigma^2$) has a central χ^2 (chi-squared) distribution with K degrees of freedom, which is denoted here by χ_K^2.

Now consider a random sample $\{X_i, i = 1, \ldots, N\}$, selected from a normal population with unknown mean μ and unknown variance σ^2. The deviation of each score from the sample mean, $X_i - \bar{X}$, has an expected value of 0 and, by a well-known algebraic argument, the sum of the squared deviations, $\Sigma(X_i - \bar{X})^2/\sigma^2 = (N - 1)s^2/\sigma^2$, has a χ_{N-1}^2 distribution. Therefore, given a random sample from a population with a normal distribution and unknown mean and variance, the confidence interval for the variance σ^2 utilizes the χ^2 distribution.

Let $\chi_{r:\alpha/2}^2$ denote the value that slices $\alpha/2$ from the right-hand tail of a chi-square distribution with r degrees of freedom. Then for a sample size of N and a sample variance of s^2, a two-sided $100(1 - \alpha)\%$ confidence interval for σ^2 is defined by

$$[(N - 1)s^2/\chi_{N-1:\alpha/2}^2, \ (N - 1)s^2/\chi_{N-1:1-\alpha/2}^2]. \quad [3.6]$$

If we prefer, we may obtain a confidence interval for the standard deviation instead of the variance simply by taking square roots of the limits on the interval in Formula 3.6. Some writers express the confidence interval for a variance using an F distribution rather than a χ^2 distribution. The relationship between the two distributions is $\chi_r^2 = rF_{r,\infty}$, where $F_{r,\infty}$ denotes an F distribution with r and infinite degrees of freedom. Thus, an alternative version of Formula 3.6 is

$$[s^2/F_{N-1,\infty:\alpha/2}, \ s^2/F_{N-1,\infty:1-\alpha/2}]. \quad [3.7]$$

Researchers should be aware that Formula 3.6 performs poorly for samples below about $N = 40$ and is sensitive to non-normality. Interested readers may wish to consult Burdick and Graybill (1992, pp. 25-26) for further information on these issues.

Example 3.4: Variance and Standard Deviation

Returning to Smithson's (2000, pp. 194-196) fictitious social facilitation study comparing measurements of 12 800-meter runners' times under two conditions (no one present vs. one person present), suppose that the variability of the social facilitation effect is of interest. The standard deviation of the differences in times is $s = 1.284$. Thus, the sample variance is $s^2 = 1.649$. The required chi-square statistics are $\chi^2_{11:\alpha/2} = 19.675$ and $\chi^2_{11:1-\alpha/2} = 4.575$. From Formula 3.6, the 90% confidence interval for χ^2 is $[(11)(1.649)/19.675, (11)(1.649)/4.575] = [0.922, 3.965]$. That is, at the 90% confidence level, the plausible values for the population variance range from 0.922 to 3.965.

In some applications, a confidence interval for the standard deviation is more meaningful than one for the variance. All we need do here is take the square root of the confidence interval limits for the variance. In this example, the corresponding confidence interval for the standard deviation is $[0.960, 1.991]$. This estimate is very imprecise (and rather inappropriate) because of the small sample.

Confidence Intervals for the Ratio of Variances

Two sample variances may be compared via their ratio, that is, σ_1^2/σ_2^2. Given two independent random samples of sizes N_1 and N_2 from populations with normal distributions, $(s_1^2/\sigma_1^2)/(s_2^2/\sigma_2^2)$ has an $F_{N-1,N-2}$ distribution. A one-sided $100(1 - \alpha)$% confidence interval for σ_1^2/σ_2^2 with a lower bound is

$$[(s_1^2/s_2^2)/F_{N-1,N-2:\alpha}, \infty], \qquad [3.8]$$

where $F_{N-1,N-2:\alpha}$ is the value that slices α from the right-hand tail of the F distribution, and a one-sided interval with an upper bound is

$$[0, (s_1^2/s_2^2)/F_{N-1,N-2:1-\alpha}]. \qquad [3.9]$$

A two-sided $100(1 - \alpha)$% confidence interval for σ_1^2/σ_2^2 is therefore

$$[(s_1^2/s_2^2)/F_{N-1,N-2:\alpha/2}, (s_1^2/s_2^2)/F_{N-1,N-2:1-\alpha/2}]. \qquad [3.10]$$

TABLE 3.2
Mood Scale Scores

Control	Treatment
1	3
9	6
10	5
0	6
8	4
2	6

Balakrishnan and Ma (1990) provide a comparative review of tests of the null hypothesis $\sigma_1^2/\sigma_2^2 = 1$ when normality cannot be assumed.

The confidence intervals introduced here for variances and variance ratios have many applications, primarily in random-effects ANOVA and regression. Some of these applications will be covered in Chapter 5, but space limitations preclude anything like an exhaustive treatment. Interested readers may wish to consult Burdick and Graybill (1992) for thorough coverage of confidence intervals for random-effects (variance components) models in ANOVA.

Example 3.5: The Variance as an Experimental Outcome

As an example, Smithson (2000, p. 230) considers a fictitious treatment designed to restrict the mood swings of people suffering bipolar disorder. Table 3.2 shows artificial data from a hypothetical study in which a self-report mood scale ranging from 0 (extreme depression) to 10 (extreme elation) was used. Because mood variability is the issue here, we want to compare the variances of the treatment and control groups.

Clearly, the treatment group has less variance. In fact, $s_2^2 = 1.6$, whereas $s_1^2 = 20.0$, so the sample ratio of the variances is $s_1^2/s_2^2 = 12.5$. A researcher in this situation might use a one-sided confidence interval if he or she were interested only in the extent to which $s_1^2 > s_2^2$. If we wish a confidence level of 99%, from Formula 3.8, we have

$$[(s_1^2/s_2^2)/F_{5,5:.01}, \infty] = [12.5/10.967, \infty] = [1.140, \infty].$$

The researcher would be able to claim that 1.14 is a lower bound on plausible values for the ratio. On the other hand, Formula 3.10 would yield the following two-sided 99% confidence interval:

$$[12.5/14.939, 12.5/0.067] = [0.837, 186.744].$$

This interval does not even exclude 1.

It is worth noting in this context that the traditional F test result would have us reporting $F_{5,5} = 12.5$, $p = .0075$. If we used a significance criterion of $\alpha = .01$, because the F test is a one-tailed test it would be associated with the one-sided 99% confidence interval, which does exclude 1. Nevertheless, despite this "highly significant" result, the small samples result in a one-sided interval whose lower limit is rather close to 1 and a wide two-sided confidence interval. Both of these are symptomatic of an imprecise estimate, but they also illustrate the trade-off in the choice between a one-sided and a two-sided confidence interval, especially where a one-tailed significance test traditionally is used. The one-sided $100(1 - \alpha)\%$ interval will never contradict the test, whereas a two-sided interval will require a confidence level of $100(1 - 2\alpha)\%$ to be consistent with the one-tailed test (e.g., we would need a 98% two-sided interval in our example). Moreover, the one-sided interval is the natural counterpart to a power calculation in this situation. However, the two-sided interval provides an appealing assessment of the precision of the estimate that the one-sided interval cannot.

Example 3.6: Comparing Variances in Heights

Given two random samples of women from Country A ($N_A = 70$) and Country B ($N_B = 90$), with standard deviations in height $s_A = 26$ cm and $s_B = 17$ cm, we want to compare the variability in women's heights for the two countries. Suppose we decide to construct the 90% confidence interval for the ratio of the variances. To start with, the sample F statistic is $F_{69,89} = 26^2/17^2 = 2.3391$ ($p < .00001$). For the two-sided confidence interval, from Formula 3.10 we get

$$[2.3391/F_{69,89:.05}, \; 2.3391/F_{69,89:.95}] = [2.3391/1.4481, \; 2.3391/0.6834]$$
$$= [1.615, 3.423].$$

Thus, although the F statistic might be regarded as "highly significant," the confidence interval is rather less impressive but more realistic. It informs us that on the basis of sample sizes of 70 and 90, we cannot resolve the ratio of the variances any more precisely than by a factor of about 2.

The Transformation Principle

Transforming a statistic may enable researchers to compute a confidence interval for the transformed version that may then be transformed back into a confidence interval for the original statistic (Cox & Hinkley, 1974; Kendall & Stuart, 1979). The crucial idea here is that confidence intervals for a parameter yield corresponding confidence intervals for any monotonic transformation of that parameter. The most common application of this "transformation principle" is in situations where a transformed statistic has a known tractable limiting distribution. Popular examples include the simple correlation coefficient and the odds ratio.

Approximate Confidence Interval for the Simple Correlation Coefficient

We have already seen that a confidence interval for the variance yields one for the standard deviation, simply by taking the square root of the limits for the variance confidence interval. Another popular example of the transformation principle in action is Fisher's transformation of R, the simple correlation coefficient, to construct a confidence interval for the population correlation. The sampling distribution for R is not a simple one, and Fisher's transformation converts R into a statistic that is approximately normally distributed. This transformation is sometimes known as Fisher's z, but to avoid confusion it will be denoted here by R':

$$R' = \frac{1}{2}\ln\left[\frac{1+R}{1-R}\right]$$ [3.11]

where "ln" stands for the natural logarithm. It turns out that the standard error of R', is approximated by

$$s_{R'} = \frac{1}{\sqrt{N-3}.}$$ [3.12]

For a confidence level of $100(1 - \alpha)$, the half-width of the interval is $w = (z_\alpha/2)(s_{R'})$, so the confidence interval is $[R' - w, R' + w]$. To convert this confidence interval back to an interval for the correlation coefficient itself, we need the inverse of the transformation we used originally:

$$R = \frac{e^{2R'} - 1}{e^{2R'} + 1}.$$ [3.13]

This transformation may be applied to the lower and upper limits of the interval $[R' - w, R' + w]$, thereby converting them into lower and upper limits on a confidence interval for simple R. Readers should bear in mind that this approximation can be poor for small samples, values of R near -1 or $+1$, and severe departures from bivariate normality.

Example 3.7: Correlation Between Two Raters

Consider a sample correlation between mothers' and fathers' aggressiveness ratings of their children in the Child Behavior Check-List (this example is detailed in Smithson, 2000, pp. 278-281). The sample value for R is .685, with a sample size of $N = 178$. Formula 3.11 gives us

$$R' = \frac{1}{2} \ln \left[\frac{1 + R}{1 - R} \right] = 0.5 \ln \left[\frac{1.685}{0.315} \right] = 0.8385.$$

From Formula 3.12, the standard error of R' is

$$s_{R'} = \frac{1}{\sqrt{N - 3}} = 0.0756.$$

Given a desired confidence level of 95%, $z_{.025} = 1.96$ and the half-width is $w = (z_{\alpha/2})(s_{R'}) = 0.1482$. Our confidence interval for the transformed R' is $[R' - w, R' + w] = [0.6903, 0.9866]$.

If we apply the inverse transformation formula 3.13 to convert this interval back to an interval for simple R, we get $[0.5982, 0.7559]$. Note that the confidence interval for simple R is usually not symmetrical around R, even though the transformed interval is symmetrical around R'. In our example, $R = 0.685$ is closer to the upper limit of 0.7559 than it is to the lower limit of 0.5982.

Approximate Confidence Interval for the Odds Ratio

Another widely used statistic (especially in health-related research) is the odds ratio, for which the transformation principle may be used in constructing an approximate confidence interval. Table 3.3 displays a 2 × 2 contingency table, with N_{ij} referring to the number of cases in the ith category on X and the jth category on Y. Now, the odds that someone will be in Category 1 on Y given that he or she is in Category 1 on X is

$$\text{Odds}(Y = 1 | X = 1) = N_{11} / N_{12}.$$

Likewise, the odds of being in Category 1 on Y for someone in Category 2 on X is

$$\text{Odds}(Y = 1 | X = 2) = N_{21} / N_{22}.$$

TABLE 3.3

A 2 × 2 Contingency Table

X	Y 1	2	Total
1	N_{11}	N_{12}	N_{1+}
2	N_{21}	N_{22}	N_{2+}
Total	N_{+1}	N_{+2}	N_{++}

The odds ratio (of the first odds to the second) is

$$\Omega = \frac{\text{Odds}(Y = 1 | X = 1)}{\text{Odds}(Y = 1 | X = 2)} = \frac{N_{11}/N_{12}}{N_{21}/N_{22}}. \qquad [3.14]$$

If we denote the sample estimate of the odds ratio by W (distinguishing it from the population odds ratio Ω), then for moderately large N_{++}, an approximate confidence interval for $\ln(\Omega)$ has the form $[\ln(W) - w, \ln(W) + w]$, where $w = (z_{\alpha/2})(s_{err})$ and the standard error of $\ln(W)$ is

$$s_{err} = \sqrt{1/N_{11} + 1/N_{12} + 1/N_{21} + 1/N_{22}}. \qquad [3.15]$$

To convert our confidence interval for $\ln(\Omega)$ into one for Ω, the inverse transformation is just $e^{\ln(\Omega)}$ applied to the lower and upper limits. Agresti (1990, p. 54) and others note that replacing N_{ij} with $N_{ij} + 0.5$ in [3.14] and [3.15] results in better-behaved estimators, particularly if any of the N_{ij} are small.

Example 3.8: Gender Predicts Odds of Becoming a Specialist

The data shown in Table 3.4 are taken from a study by Gerrity, Earp, and DeVellis (1992) of a sample of American physicians. A traditional approach to the question of whether males are more likely than females to become specialists would be to conduct a chi-square test of independence and then perhaps use an odds ratio to evaluate the strength of the relationship between gender and the tendency to become a specialist or a general practitioner. A confidence interval for the odds ratio can provide all this information and more.

TABLE 3.4

Gender and Preference for Specialist vs. General Medical Practice

Gender	Specialists	General Practitioners
Male	164	174
Female	12	39

Source: Data are from Gerrity, Earp, and DeVellis (1992).

From Formula 3.14, the odds ratio for these data is

$$W = \frac{\text{Odds(Spec.|Male)}}{\text{Odds(Spec.|Female)}} = \frac{164/174}{12/39} = \frac{0.9425}{0.3077} = 3.063.$$

A 95% confidence interval for $\ln(\Omega)$ has the form

$$[\ln(3.063) - w, \ln(3.063) + w],$$

where

$$w = (z_{\alpha/2})(s_{err}) = (1.96)(s_{err}),$$

and, from Formula 3.15,

$$s_{err} = \sqrt{1/N_{11} + 1/N_{12} + 1/N_{21} + 1/N_{22}}$$

$$= \sqrt{1/174 + 1/164 + 1/12 + 1/39} = 0.3476,$$

so $w = (1.96)(0.3476) = 0.6813$.

Because $\ln(3.063) = 1.1195$, our confidence interval is [0.4382, 1.8008].

. Finally, we raise e to the power of the lower and upper limits to get a 95% confidence interval for Ω of [1.550, 6.054]. As in the correlation example, this interval is not symmetrical around $W = 3.063$. Moreover, once again owing to the modest sample size, the odds ratio estimate is not very precise, even though the sample odds ratio is reasonably large and the trend in the table is quite clear.

4. NONCENTRAL CONFIDENCE INTERVALS FOR STANDARDIZED EFFECT SIZES

Noncentral Distributions

The most commonly used noncentral distributions in constructing confidence intervals are the noncentral t, F, and χ^2 distributions. The noncentral distributions describe reality when the null hypothesis is false, and they play an important role in both power analysis and confidence intervals for standardized effect-size measures such as multiple R^2 and Cohen's d. Standardized effect-size measures are useful because they are "scale-free," and confidence intervals for these statistics should have wide appeal. The fact that methods for constructing such intervals are missing from modern textbooks is attributable to the abandonment of these noncentral distributions in the 1950s and 1960s as a result of insufficient computing power needed to employ them.

These three noncentral distributions are closely interrelated for much the same reason that their central counterparts are. We will begin with a characterization of the noncentral χ^2 distribution because of the role it plays in the noncentral t and F distributions.

Noncentral χ^2 Distribution

Suppose we have a collection of K independent normally distributed random variables $\{Y_i, i = 1, \ldots, K\}$ whose means are all 0 and whose variances are all σ^2. As mentioned earlier, the sum of squares of these variables divided by σ^2 (i.e., $\Sigma Y_i^2/\sigma^2$) has a central χ^2 distribution with K degrees of freedom. This χ^2 variable will have a mean of K.

Now suppose the means of the K random variables are not all 0, and instead that each Y_i has its own mean μ_i. Then $\Sigma Y_i^2/\sigma^2$ has a noncentral χ^2 distribution with K degrees of freedom and a noncentrality parameter $\Delta = \Sigma \mu_i^2/\sigma^2$. The standard notation for a noncentral χ^2 with $df = K$ and noncentrality parameter $= \Delta$ is $\chi^2_{K,\Delta}$. This χ^2 variable will have a mean of $K + \Delta$.

In using Δ to denote the noncentrality parameter, I am following Cumming and Finch (2001, p. 547) instead of the more standard notation δ, for the same reasons they give (but which need not concern us here). One useful interpretation of Δ is that it is the *sum of squared standardized effects* or deviations from 0. This interpretation may be directly applied to a fixed-effects ANOVA model or a chi-square statistic for contingency table analysis. The crucial point is that $\Sigma Y_i^2/\sigma^2$ has a central χ^2 distribution *only* when the null hypothesis is true.

Noncentral F Distribution

The central F distribution is the ratio of two independent central χ^2 variables and their respective degrees of freedom. A test for the equality of two variances, for example, uses the F *ratio*, conventionally defined by $F = s_1^2/s_2^2$. What is usually omitted from this definition is the fact that the ratio really uses s_1^2/σ_1^2 and s_2^2/σ_2^2, with the null hypothesis assumption that $\sigma_1^2 = \sigma_2^2$ so that they cancel each other out in the F ratio.

By reintroducing σ_1^2 and σ_2^2, we may set up a noncentral F distribution in which $\sigma_1^2 \neq \sigma_2^2$. A simple way to do this is to set $\sigma_2^2 = 1$ so that σ_1^2 is considered as a multiple of σ_2^2. Thus, we put $F = (s_1^2/\sigma_1^2)/s_2^2$. Then our noncentral F distribution involves the ratio of a noncentral $\chi^2_{df1,\Delta}$ variable and a central χ^2_{df2} variable. As before, Δ is the sum of squared standardized effects and can be directly linked to standardized effect-size measures such as multiple R^2. Indeed, it takes only one more step to apply this setup to a fixed-effects ANOVA, as Steiger and Fouladi (1997) do. Consider a one-way experimental design with K conditions and n observations in each condition. Let α_i be the treatment effect associated with the ith condition and σ^2 be the error variance. Then the F statistic with $K - 1$ and $K(n - 1)$ degrees of freedom has a noncentrality parameter $\Delta = n\Sigma\alpha_i^2/\sigma^2$, so that Δ/n is the sum of squared standardized effects.

Noncentral t Distribution

The central t distribution is a simple function of the ratio of two random variables (say, W and V). W has a normal distribution with a mean of 0 and standard deviation of 1. V has a central χ^2_r distribution. If we put

$$t = W/(V/r)^{1/2},$$ [4.1]

then t has a central t distribution with r degrees of freedom.

In its most common application, t is the sample mean divided by the sample standard error. It helps to decompose this conventional definition into W and V and thereby set the stage for defining a noncentral t distribution. If we rewrite the one-sample t statistic as

$$t = \frac{(\bar{X} - \mu_0)/(\sigma/\sqrt{N})}{s/\sigma}$$ [4.2]

where μ_0 is a hypothetical value for the population mean, then we can see that W is the numerator of t and

$$V = SS/\sigma^2$$

where SS is the sum of squares for X.

The assumption that gives this t statistic a central t distribution is that W has a mean of 0 (i.e., the hypothesis concerning the mean is correct). If W has a mean of $(\mu - \mu_0)/(\sigma/\sqrt{N}) \neq 0$, then t has a noncentral t distribution with a noncentrality parameter

$$\Delta = (\mu - \mu_0)/(\sigma/\sqrt{N}), \qquad [4.3]$$

which is directly related to the standardized effect size. Note that this noncentrality parameter is not based on a sum of squares as in the previous two distributions. The relationship between the noncentral t and noncentral F will be addressed later. For now, it is worth noting that t is the sample estimate of Δ.

The standardized effect size in this context is Cohen's d, which may be thought of as a mean difference standardized by a relevant standard deviation. We will denote the population parameter Cohen's d by δ. In the one-sample case here, it is defined as

$$\delta = (\mu - \mu_0)/\sigma = \Delta/\sqrt{N}. \qquad [4.4]$$

Computing Noncentral Confidence Intervals

Inversion Confidence Interval Principle

Confidence intervals based on noncentral distributions depend on the "inversion confidence interval principle" as explicated by Steiger and Fouladi (1997, pp. 237-239). The main idea is to use the observed value of a test statistic (e.g., an observed multiple R^2) to initiate the search for the lower and upper limits to a $1 - \alpha$ confidence interval for the noncentrality parameter, which is then converted into a confidence interval for an effect size. That is, we move through three stages:

$$\text{test statistic} \rightarrow \text{CI for noncentrality parameter}$$
$$\rightarrow \text{CI for effect-size parameter.}$$

As long as the effect-size parameter is a monotonic function of the noncentrality parameter, then a confidence interval for the noncentrality parameter always can be transformed into a confidence interval for the effect-size parameter.

Example 4.1: Cohen's d for One Sample

Let us return to the repeated-measures t test from Smithson (2000, pp. 194-196). The fictitious social facilitation study presented there involves measurements of twelve 800-meter runners' times under two conditions (no one present vs. one person present). The mean difference is 0.983 seconds (faster for the someone-present condition), and we have already seen that the 95% confidence interval is [0.983 − 0.817, 0.983 + 0.817] = [0.166, 1.800]. Because we have an agreed-upon scale here (seconds), this is sufficient for many research purposes.

However, if we wanted to compare this social facilitation effect with that for some other event (a foot race over another distance, or a swimming or cycling race), then we would also want a "dimensionless" effect-size measure such as Cohen's d. The sample Cohen's d is

$$d = 0.983/1.284 = 0.766.$$

If we want to put a 95% CI around d, then we cannot simply divide the limits in the interval around the mean by the sample standard deviation and use that as an interval for Cohen's d. The confidence interval for Cohen's d relies on the sampling distribution for μ/σ, which has two sources of sampling variation rather than just one.

Recall that the observed $t_{11} = 0.983/(1.284/\sqrt{12}) = 2.65$. This also gives us a sample estimate of the noncentrality parameter. The next step is along the lines of the binomial distribution approach to a confidence interval for the proportion. At this juncture, we require a computer routine to home in on a lower limit Δ_L, corresponding to the noncentral t distribution for which $t_{11} = 2.65$ cuts off $\alpha/2$ of its upper tail, and an upper limit for Δ_U corresponding to the noncentral t distribution for which $t_{11} = 2.65$ cuts off $\alpha/2$ of its lower tail. SPSS, SAS, SPlus, and R routines for computing confidence intervals and power for the noncentrality parameter and related statistics for the noncentral t distribution are available free of charge from the author via his Web page (www.anu.edu.au/psychology/staff/mike/Index.html).

The resulting 95% confidence interval around the noncentrality parameter Δ is [0.3604, 4.8494]. That is, for a noncentral t distribution with a noncentrality parameter of 0.3604, we would observe a sample t_{11} of greater than 2.65 2.5% of the time, and for a noncentral t distribution with a noncentrality parameter of 4.8494, we would observe a sample t_{11} of less than 2.65 2.5% of the time. Figure 4.1 displays these noncentral t distributions in the two-distribution representation of a confidence interval analogous to Figure 2.3.

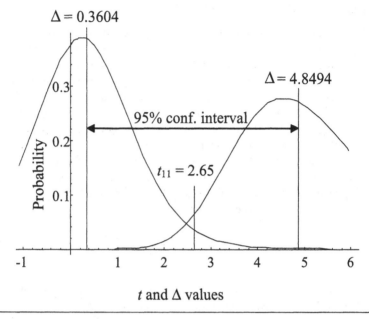

Figure 4.1   Noncentral t Distributions for 95% CI Limits for Δ

Now we turn to the third step, converting the confidence interval for the noncentrality parameter into one for the effect-size parameter. From Formula 4.4, we know that δ is $\Delta/\sqrt{N}$, and we immediately have a 95% CI for δ, namely [0.1040, 1.3999]. According to Cohen's (1988) effect-size guidelines, although $d = 0.766$ is a moderately large effect, our CI tells us that the effect size plausibly could range from quite small (0.1040) to very large (1.3999).

Example 4.2: Cohen's d *for Two Samples*

In Example 3.3, we constructed a confidence interval for the difference in mean likelihood reported by Windschitl and Wells (1998) in a two-groups experiment. Briefly recapitulating, the mean likelihood rating for the 30-7-5-5-3 condition was $\bar{X}_1 = 13.9$, and the mean for the 30-20 condition was $\bar{X}_2 = 11.9$, both on a 21-point scale. The 95% confidence interval for the difference between these means was [0.855, 3.145].

In addition to the $t_{72} = 4.00$ ($p < .001$) result, the authors also report that Cohen's $d = 0.93$. By Cohen's standards, this is a large effect. We will construct a two-sided 95% confidence interval for d. For the population Cohen's d for two samples,

$$\delta = (\mu_1 - \mu_2)/\sigma_p, \qquad [4.5]$$

where σ_p is the population pooled standard deviation corresponding to the sample estimate defined in Formula 3.4. This quantity has a noncentral t distribution with a parameter Δ, defined by

$$\Delta = \delta[N_1 N_2/(N_1 + N_2)]^{1/2}, \qquad [4.6]$$

where N_1 and N_2 are the sizes of the two samples. So

$$\delta = \Delta/[N_1 N_2/(N_1 + N_2)]^{1/2}, \qquad [4.7]$$

which is analogous to the one-sample case.

Using the same computational routine as in Example 4.1, we obtain a 95% confidence interval around the noncentrality parameter Δ of [1.922, 6.053]. Then, we convert the confidence interval for the noncentrality parameter into one for the effect-size parameter. From Formula 4.7, we have

$$\delta_L = \Delta_L/[N_1 N_2/(N_1 + N_2)]^{1/2} = 1.922/4.301 = 0.447, \text{ and}$$

$$\delta_U = \Delta_U/[N_1 N_2/(N_1 + N_2)]^{1/2} = 6.053/4.301 = 1.407,$$

so the confidence interval for δ is [0.447, 1.407]. We may therefore conclude that the true effect-size is somewhere between medium and very large (based on Cohen's benchmarks), or that the means are somewhere between 0.45 and 1.41 pooled standard deviations apart.

Note that this outcome gives a rather different impression from the apparently small difference between the means relative to the scale-range that was found in Example 3.3. Again, whether the confidence interval for the difference between the means or for Cohen's d is superior depends on the researchers' purpose and the meaningfulness of the scale from which the means were obtained. In this example, Windschitl and Wells (1998) use a different version of their "verbal uncertainty" scale in the first two studies in the same article, so it is unlikely that this particular scale can be used effectively for replications or even comparisons with the two earlier studies. The Cohen's d measure is more useful here.

Example 4.3: Multiple R^2 for Regression

Now we will illustrate the same procedure involving the noncentral F distribution for a fixed-score regression model, by constructing a confidence interval for multiple R^2. Fisher's normalizing transformation for a simple R will not work for multiple R, necessitating the approach given here. Let u denote the number of predictors in a regression model (i.e., the numerator degrees of freedom for the corresponding F statistic), N be the sample size, and

$$v = N - u - 1 \text{ (i.e., denominator degrees of freedom).}$$

The test statistic is

$$F_{u,v} = f^2(v/u), \qquad [4.8]$$

where $f^2 = R^2/(1 - R^2)$.

In a fixed-score regression model, in which we regard the scores on the independent variables as not subject to sampling error, when the null hypothesis is not true, then $F_{u,v}$ has a noncentral F distribution with the noncentrality parameter Δ that can be estimated by

$$\Delta = f^2(u + v + 1) = (u/v)[F_{u,v}](u + v + 1). \qquad [4.9]$$

Formula 4.9 enables us to construct a confidence interval from an F value reported in a research report, because all that is needed are the $F_{u,v}$ value and u and v.

To construct a two-sided $100(1 - \alpha)\%$ confidence interval for Δ, given the sample $F_{u,v}$, we again require a computer routine to home in on a lower limit Δ_L, corresponding to the noncentral F distribution for which $F_{u,v}$ cuts off $\alpha/2$ of its upper tail, and an upper limit for Δ_U corresponding to the noncentral F distribution for which $F_{u,v}$ cuts off $\alpha/2$ of its lower tail. Once we have obtained a confidence interval for Δ, we can convert it into one for multiple R^2 via

$$\text{lower } R^2 = \Delta_L/(\Delta_L + u + v + 1) \text{ and}$$
$$\text{upper } R^2 = \Delta_U/(\Delta_U + u + v + 1). \qquad [4.10]$$

For a one-sided confidence interval, we would just find the lower limit Δ_L, corresponding to the noncentral F distribution for which $F_{u,v}$ cuts off α of its upper tail, and compute the lower R^2 from Formula 4.10.

Tabachnick and Fidell's (1996, pp. 174-193) regression example has number of visits to health professionals (timedrs) being predicted by measures of mental health (menheal), physical health (phyheal), and stress-level (stress). They take the log of timedrs + 1 (logtime) and physical health (logph) and the square root of stress (sqrtstr) to provide variables whose distributions are more suitable for linear regression. The sample size is $N = 465$ and $u = 3$, so $v = N - u - 1 = 461$. Their SPSS output yields $F_{3,461} = 92.901$ and $R^2 = .3768$. If we assume fixed-score regressors, then a confidence interval for R^2 may be obtained via the methods presented above. For example, the two-sided 90% confidence interval for the noncentrality parameter is [217.3965, 342.5157], which via Formula 4.10 translates to an interval for R^2 of [.3186, .4242]. Note that even with a sample as large as 465, the 90% CI for R^2 still has a width of about 0.1.

Example 4.4: Confidence Interval for Cramer's V

For a two-way R × C contingency table, several measures of association and/or effect size are available. One member of a class of such measures based on the Pearson chi-square statistic is Cramer's V. We can exploit the asymptotic noncentral χ^2 distribution of the Pearson chi-square statistic to construct a confidence interval for V as well as providing it with an interpretation that enhances its utility as an effect-size measure.

The Pearsonian chi-square statistic can be interpreted as the sum of squared standardized effects. Given observed frequencies o_k, we compare them with expected frequencies e_k by constructing standardized residuals

$$Z_k = \sqrt{(o_k - e_k)^2/e_k} , \qquad [4.11]$$

and then squaring and summing them. As the sample size increases, the standardized residual's distribution approximates a standardized normal distribution (i.e., with mean of 0 and standard deviation of 1).

This argument enables us to interpret the noncentrality parameter Δ of a noncentral $\chi^2_{r,\Delta}$ statistic in terms of the sum of squared standardized deviations of the o_k from the e_k. We can claim that $\Sigma Z_k^2/\sigma^2$ has a noncentral χ^2 distribution with $r = K - 1$ degrees of freedom and a noncentrality parameter $\Delta = \Sigma \zeta_k^2/\sigma^2$, where the ζ_k are the population values of the Z_k. Because these are standardized residuals, $\sigma^2 = 1$, and we are left with just $\Delta = \Sigma \zeta_k^2$.

A major limitation of χ^2 as an effect-size measure is that it increases with sample size and degrees of freedom. After all, the mean of a $\chi^2_{r,\Delta}$ variable is $r + \Delta$, and Δ is a multiple of N. It would be desirable to have a standardized effect-size measure that is not affected by r or N. The tactic we will adopt here is along the lines recommended by some authors for a fixed-effects one-way ANOVA, namely taking $\chi^2/[N(K-1)]$ to be the average squared standardized effect, where $K - 1$ is the smaller of $C - 1$ and $R - 1$. The root mean squared standardized effect (RMSSE) is Cramer's V statistic:

$$V = \sqrt{\frac{\chi^2}{N(K-1)}} \qquad [4.12]$$

A confidence interval for Δ can be transformed into one for V by the following transformation:

$$V_L = \sqrt{(\Delta_L + r)/N(K-1)}, \text{ and}$$

$$V_U = \sqrt{(\Delta_U + r)/N(K-1)}, \qquad [4.13]$$

where $r = (C-1)(R-1)$, which is the df for the chi-square statistic.

Conover (1980, pp. 156-157) presents a fairly typical example of a chi-square test of independence that could be supplemented via Cramer's V and a confidence interval. In my adaptation of his example, students randomly selected from private and public secondary schools are given standardized achievement tests, and their scores are grouped in four ranked groups, as shown in Table 4.1.

The observed chi-square value is $\chi_3^2 = 17.30$ ($df = (R-1)(C-1) = 3$, and $p = .001$). On its own, the chi-square statistic looks rather large and very significant. In this case, $K - 1 = 1$, so from Formula 4.12 the sample estimate of $V = (17.30/128)^{1/2} = .368$. This is a medium-sized effect by Cohen's standards. Using a computer routine such as SPSS, SAS, SPlus, or R (routines are available free of charge from the author via his Web page at www.anu.edu.au/psychology/staff/mike/Index.html to find lower and upper limits for a 95% confidence interval on Δ yields [3.392, 35.014]. Using Formula 4.13, we can convert this interval into a 95% interval for V:

$$V_L = [(3.392 + 3)/128]^{1/2} = .223 \text{ and}$$

$$V_U = [(35.014 + 3)/128]^{1/2} = .545.$$

The confidence interval [.223, .545] for V provides a somewhat more sober picture, telling us that the estimated effect size is moderate but the

TABLE 4.1
Test Scores

Group	A	B	C	D	Totals
Private	6	14	17	9	46
Public	30	32	17	3	82
Totals	36	46	34	12	128

confidence interval is rather wide (because of the relatively modest sample size of 128).

In this chapter, we have explored material that is unfamiliar to a great many researchers in the human sciences, although its theoretical basis was developed many years ago. It is very rare indeed to see a confidence interval for multiple R^2 or Cohen's d reported in the research literature, and it is not to be found in modern standard textbooks. It appears that the noncentral t, F, and chi-square distributions were neglected initially because without adequate computing power, they were not usable. Later, they were forgotten because of their previous neglect. Their rediscovery in the 1990s and the availability of user-friendly software for implementing them should encourage researchers to make much greater use of them. In addition to the material in this monograph, detailed expositions of their applications in specific techniques have been published recently (Cumming & Finch, 2001; Fidler & Thompson, 2001; Smithson, 2001; Steiger & Fouladi, 1997).

5. APPLICATIONS IN ANOVA AND REGRESSION

This chapter applies and extends the materials developed in Chapters 2-4 in the setting of ANOVA and regression, two mainstays of the general linear model. We begin with a brief coverage of fixed- and random-effects ANOVA models. Fixed-effects models are more easily dealt with, and the treatment given here is more extensive than that for random-effects models. Because a priori contrasts may be employed in preference to ANOVA when the researcher has specific comparisons to make, the third section of the chapter surveys methods of dealing with contrasts and relevant effect-size measures.

The next section extends the method for constructing a confidence interval for multiple R^2 to partial and semi-partial (part) correlations. Readers are alerted to the distinction between fixed-score and random-score regression models; most of the methods discussed in this chapter pertain to fixed-score models only. Techniques included are hierarchical regression, MANOVA, and setwise regression. The material on partial and semi-partial (or part) correlations is then related to confidence intervals for individual regression coefficients and related parameters. The chapter ends with an overview of noncentral confidence interval estimation in goodness-of-fit indices for structural equation models.

Fixed-Effects Versus Random-Effects Models

In experimental studies, the usual distinction that pertains to random error is between *fixed-effects* and *random-effects* designs. The definitions of these are familiar, but I include them here because later we will need to disentangle this distinction from another, less familiar one that sounds quite similar. As Howell (1997, p. 330) puts it:

> In a fixed model, the treatment levels are deliberately selected and would remain constant from one replication to another. . . . In a random model, treatment levels are obtained by a random process and would be expected to vary across replications.

The importance of this distinction lies in the fact that random-effects models introduce another source of sampling error in addition to randomized assignment, namely the selection of treatment levels. That, in turn, increases error variance in the dependent variable (DV) and requires a different error model.

The fixed-effects model can be defined as follows:

$$Y_{ij\ldots m} = \mu + \alpha_j + \beta_k + \gamma_{jk} + \cdots + e_{ij\ldots m}, \qquad [5.1]$$

where μ, α_j, β_k, γ_{jk}, etc. are *constant* (fixed-effect) *parameters* and $e_{ij\ldots m}$ is an unobservable random variable with a mean of 0 and variance σ_e^2. The main objective is to make inferences about the parameters. Noncentral F distributions are applicable here because the fixed-effects model incorporates sums of effects and therefore, by extension, sums of squared standardized effects.

The random-effects model, on the other hand, usually is characterized in terms of variance components:

$$Y_{ij\ldots m} = \mu + A_j + B_k + C_{jk} + \cdots + E_{ij\ldots m}, \qquad [5.2]$$

where the A_j, B_k, . . . , and $E_{ij\ldots m}$ are *random variables* with means of 0 and variances σ_A^2, σ_B^2, . . . , σ_E^2. The usual objective regarding this model is inferences about the variances or particular functions of them. Because the means of these random variables are defined as 0, central F distributions apply here.

Fixed-Effects ANOVA

We may use the methods described in Example 4.3 for constructing a confidence interval for the equivalent of multiple R^2 in any fixed-effects factorial ANOVA, because the same distribution theory holds there and the relevant effects can be measured via squared partial correlations or simple functions of them. The noncentrality parameter for the α_j (the "row") effect in Formula 5.1, for example, is

$$\Delta_\alpha = \Sigma N_j. \, (\alpha_j^2/\sigma_e^2), \qquad [5.3]$$

where $N_j.$ is the total number of cases in the jth row (i.e., summed over the K columns) and σ_e^2 is the error variance. Clearly, the noncentrality parameter is the sum of squared standardized effects. When there are equal numbers of cases in each cell (n, say), $N_j = nK$, where K is the number of columns.

The noncentrality parameter is closely related to the "correlation ratio" (cf. Fleishman, 1980):

$$\eta^2 = \sigma_\alpha^2/(\sigma_e^2 + \sigma_\alpha^2), \qquad [5.4]$$

where $\sigma_\alpha^2 = \Sigma N_j.\alpha_j^2/N$. The correlation ratio is, in fact, a squared partial correlation. We also have Fleishman's (1980) "signal-to-noise ratio" as in Formula 4.8:

$$f^2 = \sigma_\alpha^2/\sigma_e^2. \qquad [5.5]$$

Formulas 4.8 through 4.10 hold for the relationships among Δ, f^2, and the F statistic associated with the effect, $F_{u,v}$. As with R^2, the parameters Δ, f^2, and η^2 are related by

$$\eta^2 = f^2/(1 + f^2) = \Delta/(\Delta + u + v + 1). \qquad [5.6]$$

Example 5.1: One-Way Fixed-Effects Model

Fleishman (1980) gives an example of a one-way ANOVA with five conditions and 55 cases, with $F_{4,50} = 11.2213$. The estimated f^2 is 0.8975, and the estimated η^2 is .4730. We will construct 90% confidence intervals for these parameters.

We have $u = 4$ and $v = 50$, and the 90% CI for the noncentrality parameter turns out to be [19.385, 71.5522]. From Formulas 4.10 and 5.6, we can translate that interval into one for f^2 of [0.3524, 1.3009] and one for η^2 of [.2606, .5654].

Example 5.2: Two-Way Fixed-Effects Model

We will utilize Steiger and Fouladi's (1997) Example 7, a 2×7 factorial ANOVA with four observations per cell, presented as their Table 9.3. Faced with the usual F tests and p levels, the authors point out that researchers "could be forgiven" for supposing that the row effect was the "most significant," followed by the column effect, given the row effect $F_{1,42} = 6.00$ ($p = .0186$), the column effect $F_{6,42} = 2.65$ ($p = .0285$), and the interaction effect $F_{6,42} = 2.50$ ($p = .0369$). Indeed, the row effect has the lowest significance level.

But partial η^2 values help us to avoid confusing significance level with effect size. From Formulas 4.8 and 5.6, we find $\eta_\alpha^2 = .1250$, $\eta_\beta^2 = .2746$, and $\eta_\gamma^2 = .2632$. The row effect is, in fact, the smallest effect. If we construct 90% confidence intervals for the partial η^2 coefficients, we obtain the following intervals:

row effect Δ_α interval = [0.5208, 17.1211], η_α^2 interval = [.0117, .2801];

column effect Δ_β interval = [0.8699, 27.2893], η_β^2 interval = [.0174, .3577]; and

interaction effect Δ_γ interval = [0.4621, 25.8691], η_γ^2 interval = [.0093, .3455].

The interval for the noncentrality parameter is narrowest for the row effect because the row effect has $u = 1$ and $v = 42$, whereas the column effect and interaction effect both have $u = 6$ and $v = 42$. This is what is driving the differences in the significance levels of the F tests.

Alternative Effect-Size Measures

From the foregoing material, it should be clear that partial η^2 coefficients are not the only reasonable effect-size measures that could be used in a fixed-effects ANOVA. Cohen (1988) used the signal-to-noise ratio f^2 for power analysis in regression, ANOVA, and other members of the general linear model (GLM) family because it readily subsumes so many popular effect-size measures and enabled him to economize on the construction of power tables. Because η^2 and f^2 are monotonically related, confidence intervals for them convey essentially the same information, and which one is used is largely a matter of taste.

However, Steiger and Fouladi (1997) propose normalizing the noncentrality parameter for an effect in terms of its degrees of freedom and the cell size associated with that effect, to produce a root mean squared standardized effects (RMSSE) measure that is proportional to f. For any effect ζ, the RMSSE is defined by

$$\text{RMSSE}_\zeta = \sqrt{\frac{\Delta_\zeta}{N_\zeta u_\zeta}} \qquad [5.7]$$

where N_ζ is the total number of observations in the aggregated cell means used to evaluate the effect and u_ζ is the degrees of freedom associated with that effect. Their argument for this is based on the observation that averaging effects through dividing by nK, for instance, instead of $n(K-1)$, appears to underestimate the true effects. They are reasoning along the lines of the Cramer's V statistic, which we have seen also has an RMSSE interpretation.

In their example of a 2×7 factorial ANOVA with four observations per cell, the row effect had 2 "cells" of 28 each and 1 degree of freedom, so

$$\text{RMSSE}_\alpha = \sqrt{\frac{\Delta_\alpha}{(28)(1)}} \ ;$$

whereas the column effect had 7 "cells" of size 8 and 6 degrees of freedom, so

$$\text{RMSSE}_\beta = \sqrt{\frac{\Delta_\beta}{(8)(6)}} \ ;$$

and the interaction effect had 14 cells of size 4 and 6 degrees of freedom, so

$$\text{RMSSE}_\gamma = \sqrt{\frac{\Delta_\gamma}{(4)(6)}} \ .$$

The point of this is that we should be aware that confidence intervals for RMSSEs may differ in width, whereas their counterparts for Δ, η^2, and f^2

do not. Suppose, for instance, that the column and interaction effects in this example had identical sample F ratios of 2.65. Their degrees of freedom are identical ($u = 6$ and $v = 42$), so the 90% confidence intervals for Δ_β and Δ_γ (and therefore for their η^2 and f^2 counterparts) would also be identical. Nevertheless, their RMSSE confidence intervals differ because of the different "cell" sizes, and not only in value but also in width. The 90% confidence interval for both noncentrality parameters is [0.8699, 27.2893], which yields a confidence interval for $RMSSE_\beta$ of [.1346, .7540] and for $RMSSE_\gamma$ of [.1904, 1.0663]. The latter is wider than the former.

In their interpretation of the RMSSE confidence intervals for their example, Steiger and Fouladi (1997) remark that the wider interval for the interaction effect RMSSE indicates there is "less power (and precision of estimate) for detecting interaction effects than for detecting main effects in this design" (p. 247). This is not altogether true and certainly is not the impression conveyed by the confidence intervals for Δ, η^2, and f^2. In this particular example, the RMSSE scaling produces confidence intervals of similar width for the row and column RMSSEs (unlike the intervals for Δ, η^2, and f^2) but a wider interval for the interaction effect. Moreover, the power curves for the column and interaction effects are identical by definition. Researchers must not only choose their effect-size measures with care but also exercise caution about how widely they can generalize confidence interval statements to other effect-size measures.

Random-Effects ANOVA

The random-effects model, to remind readers of Formula 5.2, is usually set up in terms of variance components:

$$Y_{ij\cdots m} = \mu + A_j + B_k + C_{jk} + \cdots + E_{ij\dots m},$$

where the A_j, B_k, . . . , and $E_{ij\dots m}$ are m jointly independent random variables with means of 0 and variances σ_A^2, σ_B^2, ... , σ_E^2. Because the means of these random variables are defined as 0, central chi-square and F distributions apply to the construction of confidence intervals for functions of these variance components. The most thoroughgoing treatment of confidence intervals for variance-component models that I am aware of is Burdick and Graybill's (1992) monograph. We will cover only elementary models and examples here; the reader who needs more should consult that work.

We begin with the one-factor model, $Y_{ij} = \mu + A_j + E_{ij}$, where A_j and E_{ij} are jointly independent random variables with means of 0 and variances σ_A^2 and σ_E^2, so that the variance of an observation is $\sigma_A^2 + \sigma_E^2$. We also assume multivariate normality and a balanced design, that is, equal numbers of cases per cell, denoted by I. Define $df_A = J - 1$, where J is the number of groups, and $df_E = J(I - 1)$. Then a $100(1 - \alpha)\%$ confidence interval for $f^2 = \sigma_A^2/\sigma_E^2$ is

$$[(L - 1)/I, (U - 1)/I], \qquad [5.8]$$

where $L = F_{dfA, dfE}/F_{dfA, dfE:\alpha/2}$ and $U = F_{dfA, dfE}/F_{dfA, dfE:1 -\alpha/2}$,

$F_{dfA, dfE:\alpha/2}$ is the value that slices $\alpha/2$ from the right tail of the $F_{dfA, dfE}$ distribution, and

$F_{dfA, dfE:1 - \alpha/2}$ is defined correspondingly.

The intraclass correlation is defined as $\rho_A^2 = \sigma_A^2/(\sigma_A^2 + \sigma_E^2) = f^2/(1 + f^2)$, and the confidence interval for it is a transformation of Formula 5.8:

$$[(L - 1)/(L - 1 + I), (U - 1)/(U - 1 + I)]. \qquad [5.9]$$

Formulas 5.8 and 5.9 readily generalize to main and interaction effects for more complex balanced factorial designs. Methods of constructing confidence intervals for ratios of linear combinations of variance components are detailed in Burdick and Graybill (1992).

An important difficulty inherent in both Formulas 5.8 and 5.9 is the possibility that either limit may be negative. Because both f^2 and ρ_A^2 are nonnegative by definition, a negative limit is a quandary. Writers on this topic are divided on what to do about this. Burdick and Graybill (1992, p. 65), for instance, recommend defining negative bounds to be zero. Scheffé (1959, pp. 229-231), on the other hand, believes that shortening the interval by redefining negative limits to be zero provides a misleading picture about the precision of the estimate. He recommends leaving the bounds as they are but reporting the values of the sample estimates s_A^2 and s_E^2. It is likely, though, that negative bounds indicate that the underlying model is wrong and the relevant variance components are not independent, as assumed.

For unbalanced designs, the confidence intervals defined in Formulas 5.8 and 5.9 are no longer accurate, because in general the numerator of the corresponding F statistic no longer has a chi-square distribution. Wald (1940) found a method for constructing exact confidence intervals for f^2 (and therefore ρ_A^2), but it requires the solution of two simultaneous nonlinear equations. Burdick, Maqsood, and Graybill (1986) demonstrate how to use simple programming to iteratively approximate solutions, and Harville and Fenech (1985) provide easily computable approximate intervals as an alternative procedure.

Example 5.3: One-Way Model

Example 5.1 presented a one-way ANOVA with five conditions and 55 cases, with $F_{4,50} = 11.2213$, and that will do here for comparative purposes. Suppose this was a random effects setup. To start with, $F_{4,50:.05} = 2.5572$ and $F_{4,50:.95} = 0.1755$. Because I = 11, from Formula 5.8 the 90% confidence interval for f^2 is

$$[((11.2213/2.5572) - 1)/11, ((11.2213/0.1755) - 1)/11]$$
$$= [0.3080, 5.7217].$$

This is considerably wider than the fixed-effects CI of [0.3524, 1.3009]. The reason is the greater amount of sampling error introduced by assuming a random-effects model.

A Priori and Post Hoc Contrasts

A Priori Contrasts

In this section, we will examine approaches for multiple comparisons involving linear combinations of means, or "linear contrasts." In general, a linear contrast is defined by

$$\Psi = \Sigma c_k \mu_k, \qquad [5.10]$$

where the c_k are the contrast weights (usually constrained to sum to 0) and the μ_k are the group (or experimental condition) means that make up the contrast. The t statistic associated with the contrast has the general form

$$t = \Psi/[\Sigma q_k c_k^2/n_k]^{1/2}, \qquad [5.11]$$

where n_k is the number of cases in the kth component of the contrast. Assuming homogeneity of variance among the groups, $q_k = MS_w$, which is the mean-squared error within groups for the entire study. The general form of a confidence interval for Ψ is

$$\Psi \pm g[\Sigma q_k c_k^2/n_k]^{1/2}, \qquad [5.12]$$

where g is defined according to the procedure being used to control confidence level.

For planned contrasts, a popular method of controlling confidence level for k simultaneous contrasts is the Bonferroni adjustment, which amounts to using a confidence level of $1 - \alpha/k$ instead of $1 - \alpha$. Thus, in Formula 5.12 $g = t_{df_w:\alpha/2k}$, where df_w is the degrees of freedom within groups. Note that this adjustment is conservative and becomes quite restrictive for values of k beyond 5 or so.

When homogeneity of variance has been violated, then the MS_w and df_w become problematic. In formula 5.11 for the t statistic and Formula 5.12 for the confidence interval, $q_k = s_k^2$, which is the sample variances for each group, and the df_g associated with it for unequal variances is

$$df_g = \frac{[\sum s_k^2 c_k^2/n_k]^2}{\sum (s_k^2 c_k^2/n_k)^2/(n_k - 1)} .$$ [5.13]

Example 5.4: Psychomotor Performance in Children

Rosenthal and Rosnow (1985, pp. 1-3, 11-12) present an instructive example in which planned contrasts reveal a trend that ANOVA fails to detect, in which the psychomotor skills of children in five age groups comprising 10 subjects each are evaluated via their performance on a video game. Table 5.1a shows the mean performance level of each age group. Table 5.1b shows the ANOVA reported in Rosenthal and Rosnow's Tables 1.1 and 1.2. The F test in this ANOVA is nonsignificant, thereby gainsaying a trend in the means that strikes the eye as obvious. The means are almost perfectly linearly related to age.

A planned contrast, on the other hand, detects this trend. If a researcher hypothesized that the mean performance would increase with age level, then appropriate contrast weights would be $\{c_k\} = \{-2, -1, 0, +1, +2\}$. From formula 5.10, we have

$$\Psi = (-2 \times 25) + (-1 \times 30) + (0 \times 40) + (1 \times 50) + (2 \times 55) = 80.$$

Assuming homogeneity of variance, from Formula 5.11 we get

$$t = 80/[1{,}575(4/10 + 1/10 + 0/10 + 1/10 + 4/10)]^{1/2}$$
$$= 80/(1{,}575)^{1/2} = 2.016,$$

and for this t statistic, $p < .05$. Rosenthal and Rosnow (1985) use the equivalent F test to obtain this result.

TABLE 5.1A

Performance by Age Group

	Age Level				
	11	*12*	*13*	*14*	*15*
Mean	25	30	40	50	55

TABLE 5.1B

ANOVA FOR DATA IN TABLE 5.1A

Source	SS	df	MS	F	p
Age group	6,500	4	1,625	1.03	.40
Within	70,875	45	1,575		

The contrast is statistically significant, but what would a 95% confidence interval tell us about it? In formula 5.12, because we are performing only one contrast, k = 1, so $g = t_{45:\alpha/2} = 2.014$. The confidence interval for this contrast is therefore

$$\Psi \pm g[\mathrm{MS}_w \Sigma c_k^2/n_k]^{1/2} = 80 \pm (2.014)(1,575)^{1/2},$$

which yields the interval [0.068, 159.932]. In other words, the study has found a trend but is not able to resolve it at all precisely because of the small numbers of people in each cell.

Standardized Planned Contrasts and Effect Size

Steiger and Fouladi (1997, pp. 242-244) cover the topic of confidence intervals for standardized contrasts. This subsection owes much to their treatment.

Assuming homogeneity of variance and using the pooled standard deviation σ, we may convert Ψ in Formula 5.10 into a standardized contrast, $\Psi_s = \Psi/\sigma$. This simply re-expresses the contrast in pooled standard deviation units, which is one way of developing a scale-free effect-size measure for contrasts. The *t* statistic associated with the standardized contrast has a noncentral *t* distribution with noncentrality parameter

$$\Delta = \Psi_s/(\Sigma c_k^2/n_k)^{1/2}, \qquad [5.14]$$

so it is clear that a confidence interval for Δ gives us a corresponding interval for Ψ_s.

A more traditional approach (cf. Rosenthal & Rosnow, 1985) to assessing effect size in planned contrasts is to use a correlation coefficient $R = [t^2/(t^2 + df_w)]^{1/2}$. This is a simple correlation, and a confidence interval can be constructed for it using the Fisher transformation discussed in Chapter 3.

These two effect-size measures may give rather different impressions of the magnitude of effects. This is an area where agreed-upon benchmarks and comparisons of alternative measures have yet to be attained. For the time being, it is probably instructive to use both.

Example 5.4 Continued

The value of the observed standardized contrast from Example 5.4 is

$$\Psi_s = t(\Sigma c_k^2/n_k)^{1/2} = (2.016)(1) = 2.016.$$

The impression from this effect-size measure is of a very large effect. The 95% confidence interval for Δ is [0.0010, 4.0089]. From Formula 5.14, because in this case $\Sigma c_k^2/n_k = 1$, this is also the confidence interval for Ψ_s. This interval tells us that the effect size could be anywhere from nearly 0 to about 4 standard deviation units.

Now let us try the correlation approach. To begin with, the observed $R = [t^2/(t^2 + df_w)]^{1/2} = [4.063/(4.063 + 45)]^{1/2} = 0.288$. Note that contrary to the standardized contrast, this does not seem to be a particularly large effect.

However, we get much the same picture as with the standardized contrast regarding the precision of the estimate. Fisher's transformation as defined in equation 3.11 yields

$$R' = \tfrac{1}{2} \ln \left[\frac{1 + .288}{1 - .288} \right] = 0.296.$$

From Formula 3.12, the standard error of R' is $s_{R'} = 1/(N - 3)^{1/2} = 0.146$.

Using the normal approximation, $w = z_{\alpha/2}s_{R'} = (1.96)(0.146) = 0.286$, so the confidence interval is [.0105, .5823]. To convert this confidence interval back to an interval for the correlation coefficient itself, we need the inverse of the transformation in Formula 3.13, which results in a confidence interval for R of [.0105, .5243]. The result is that we can say only that the effect could range from very small to reasonably large.

Post Hoc Contrasts

This section is necessarily very brief and covers much the same ground as Maxwell and Delaney (1990, pp. 186-201) but in less detail. Among the large variety of post hoc contrast procedures, only some lend themselves to simultaneous confidence intervals as the Bonferroni procedure does. The Tukey's HSD, Dunnett, and Scheffé procedures permit simultaneous confidence intervals, but Fisher's Least Significant Difference, the Newman-Keuls procedure, and the Duncan method are examples of stagewise or nested approaches that do not enable confidence intervals to be computed. Tukey's procedure is used for all pairwise comparisons, Dunnett's for comparisons with a control or baseline condition, and Scheffé's for all contrasts.

Scheffé's procedure in particular is very flexible and popular, if somewhat conservative. Referring to Formula 5.12, for the Scheffé method

$$g = [(k - 1)F_{k-1, df:\alpha}]^{1/2}, \qquad [5.15]$$

where $df = df_w$ assuming equal variances and $df = df_g$ otherwise. For details on the Tukey and Dunnett procedures, the reader should consult Maxwell and Delaney (1990, pp. 199-201).

Regression: Multiple, Partial, and Semi-Partial Correlations

We have already seen how to construct a confidence interval for multiple R^2 using the noncentral F distribution. In this section, we will extend that treatment to partial and semi-partial (part) correlations, largely following the framework laid out by Cohen (1988). Before we proceed to those extensions, however, several other issues must be addressed. The next two subsections deal with the distinction between fixed- and random-score models, and with the use of adjusted and shrunken R^2.

Fixed-Score Versus Random-Score Regression Models

A distinction that may not be as familiar as fixed- versus random-effects models but is just as important is between *fixed scores* and *random scores* in regression. Fixed scores are preplanned, whereas random scores arise from random sampling. For instance, if a 2×3 factorial design with 60 subjects includes the stipulation that all cells in the design will have equal numbers of people, then both of those independent variables' scores are *fixed*, regardless of whether either of them has a fixed-effect or random-effect setup. Conversely, if the same design randomly assigns the 60 subjects to cells in such a way that the number of people in each cell is a random variable (i.e., cells could have unequal size by chance), then both independent

TABLE 5.2
Fixed-Score vs. Random-Score CIs for R^2

$F_{u,v}$	u	v	Fixed Score		Random Score	
			R^2 for L	R^2 for U	R^2 for L	R^2 for U
5.0000	4	30	.1036	.5121	.1040	.5526
8.3333	4	50	.1822	.5006	.1799	.5306
13.3333	4	80	.2364	.4881	.2323	.5105
16.6667	4	100	.2570	.4819	.2522	.5015
20.0000	4	120	.2719	.4770	.2670	.4945
23.3333	4	140	.2831	.4729	.2781	.4888
26.6667	4	160	.2919	.4694	.2869	.4840
82.5000	4	495	.3428	.4439	.3392	.4513

variables' scores are *random*, again regardless of whether they involve fixed or random effects. Fixed and random effects pertain to levels, whereas fixed and random scores pertain to the distribution of scores.

How does this issue influence the confidence intervals we use? And why is it not discussed in connection with the usual significance tests? Fixed-score models permit narrower confidence intervals for multiple R^2 than do random-score models because the random-score model takes account of sampling error in the independent variables. When the null hypothesis is true, however, the sampling distribution for R^2 does not differ, so the same F distribution can be used in either case for the conventional significance test. We still may use the fixed score-based confidence intervals for such things as multiple R^2 and regression coefficients as long as we bear in mind that they are *conditional* confidence intervals—conditional on the particular scores observed in the sample.

The noncentral F distribution cannot be used to construct confidence intervals for multiple R^2 in random-score models, but serviceable algorithms are available for constructing such intervals (cf. the "R2" program by Steiger and Fouladi, 1992, and Mendoza and Stafford, 2001). How big a difference does this make? How big is the difference between the confidence intervals for fixed- and random-score models? Table 5.2 provides an illustration, in this case for a four-predictor regression model with various sample sizes and an F value that corresponds to an observed $R^2 = 0.4$. The confidence level is .90.

The first pair of confidence interval limits, assuming fixed scores, was computed using the noncentral F (SPSS syntax available from the author), whereas the second pair, assuming random scores, was computed using the

"R2" program. The second confidence interval is clearly wider than the first, and the difference between the widths decreases only slowly with increased sample size. The difference between the confidence intervals cannot be ignored unless samples are fairly large. Researchers will have to make their own choices regarding which kind of model to use.

Adjusted and Shrunken R^2

Because the sample multiple R^2 is a positively biased estimator of its true population value, several "adjusted" or "shrunken" estimators have been proposed as substitutes. It might seem intuitively reasonable to consider confidence intervals based on these parameters. Adjusted R^2 (denoted here by R_a^2) is applicable when we want to estimate population multiple R^2 because it is less biased than sample R^2. The traditional R_a^2 is related to sample R^2 by replacing sums of squares with mean squares. Thus, whereas $R^2 = 1 - SS_e/SS_Y$, the adjusted estimate is $R_a^2 = 1 - MS_e/MS_Y$ and is related to R^2 by

$$R_a^2 = R^2 - (u/v)(1 - R^2). \qquad [5.16]$$

Although $R_a^2 < 0$ when $F_{u,v} < 1$, this outcome does not contradict the statistical significance test associated with $F_{u,v}$. It is clear that a confidence interval for the noncentrality parameter not only provides a confidence interval based on R^2 but also could give us one based on R_a^2.

However, as pointed out in Smithson (2001), there are arguments against using an interval based on R_a^2. These arguments will be repeated here. To begin with, the procedure presented earlier constructs a confidence interval for the noncentrality parameter of the F distribution, which in turn is the sum of squared standardized effects. Both sample R^2 and R_a^2 are merely alternative monotonic functions of this parameter and u and v.

We must also bear in mind that unbiasedness for a pointwise estimator does not guarantee other properties pertaining to the confidence interval. First, unbiasedness for confidence intervals refers to the accuracy of coverage (e.g., do 95% of confidence intervals really include the true parameter value?). Second, one estimator may be less biased than another in the pointwise estimation sense but produce a wider confidence interval than the other. It is worth recalling that a confidence interval is not necessarily an interval "around" a pointwise estimator.

A major problem with an interval based on R_a^2 is that it can give a negative lower limit even when the lower limit on the noncentrality parameter is larger than 0. To see this, consider a one-sided confidence interval. Denoting the critical value for the F test at any α level by $F_{u,v} = C_\alpha$, when the observed $F_{u,v} < C_\alpha$, then by definition the lower limit of a confidence interval for Δ (and therefore for multiple R^2) must be 0. That turns out to be true for a confidence interval based on sample R^2, but an interval based on R_a^2 gives a negative lower limit. In fact, such an interval can give negative lower limits (thereby including 0) when the true lower limit on the noncentrality parameter is larger than 0, thereby flatly contradicting the F test. A bit of algebra suffices to show that the lower limit of the confidence interval based on R_a^2 will fall below zero whenever $(u/v)(u + v + 1) > \Delta_L$, where Δ_L is the lower limit on the confidence interval for Δ.

If we turn to two-sided confidence intervals, a two-sided 90% confidence interval for Δ will have the same lower bound as a one-sided 95% interval, but the two-sided interval displays all the hypothetical values for Δ that we could associate with a two-tailed F test. Again, sample R^2 is consistent with this two-tailed test in ways that R_a^2 is not. A two-sided interval based on R_a^2 not only has lower limits than one based on R^2 but is also wider by a factor of $1 + u/v$.

Thus, for anyone who wishes confidence intervals to be consistent with the associated significance tests, R_a^2 is not appropriate for characterizing confidence intervals associated with squared multiple correlation. It is, of course, still a less biased pointwise estimator than sample R^2 and also should play a role in designing studies when confidence interval width is a consideration. This application will be taken up in Chapter 7.

Shrunken R^2 is applicable when we ask how well a regression model based on our sample will predict the dependent variable for a new random sample from the same population (cf. Darlington, 1990). Thus, there is a unique "true value" of shrunken R^2 for each sample. The most common method of estimating shrunken R^2 is by cross-validation, whereby the regression coefficients in one sample are used to generate predicted Y values in a second sample. The resulting shrunken R^2 is just the squared correlation between the predicted and actual Y values in the second sample, so the usual normal approximation via the Fisher's z transformation may be used with this estimate to obtain a confidence interval for it. Other methods of obtaining shrunken R^2 do not lend themselves as easily to the construction of confidence intervals, with the exception of bootstrap methods (cf. Diaconis & Efron, 1983; Efron & Tibshirani, 1993).

Semi-Partial and Partial R²

In this subsection, we extend the methods developed in Example 4.3 to constructing confidence intervals related to semi-partial and partial multiple correlation coefficients, popular effect-size measures of the contribution that one set of predictors (set B, say) provides in predicting a dependent variable when another set (A) has been taken into account. Following Cohen's (1988) notation, let y = number of predictors in set A, u = number of predictors in set B, and define $v = N - u - y - 1$. The squared semi-partial (or part) correlation is defined by $R^2_{Y.A,B} - R^2_{Y.A}$, namely the difference between the squared multiple correlation when both sets A and B are included in the model and when only set A is included (often referred to as R^2-change). The usual way of assessing whether set B has made a statistically significant contribution to predicting Y is to test whether the R^2-change is 0 via an F test. The relevant F statistic may be written as $F_{u,v} = f^2(v/u)$ as in equation 5.15, as before, but this time the signal-to-noise ratio f^2 is a function of the squared *partial* correlation:

$$f^2 = R^2_{YB.A}/(1 - R^2_{YB.A}), \qquad [5.17]$$

where the squared partial correlation is defined as

$$R^2_{YB.A} = (R^2_{Y.A,B} - R^2_{Y.A})/(1 - R^2_{Y.A}). \qquad [5.18]$$

In a fixed-score regression model, when the null hypothesis regarding the contribution of set B is not true, then $f^2(v/u)$ has a noncentral F distribution with noncentrality parameter $\Delta = f^2(u + v + 1)$ as before, and we also have

$$R^2_{YB.A} = f^2/(1 + f^2) = \Delta/(\Delta + u + v + 1). \qquad [5.19]$$

Thus, it is straightforward to construct a confidence interval for f^2 and partial R^2 but not for the squared semi-partial correlation.

Because the squared partial correlation is just the squared semi-partial correlation divided by $1 - R^2_{Y.A}$, for most purposes (including meta-analysis) f^2 and $R^2_{YB.A}$ provide everything needed. The difference between the partial and semi-partial correlation becomes important when $1 - R^2_{Y.A}$ is well below 1, which occurs when the set A predictors already explain a substantial proportion of the variance in the dependent variable. There are also issues concerning the robustness of partial correlations, but these are usually dealt with in standard textbooks on regression (e.g., Darlington, 1990, pp. 134-136).

Example 5.5: Assessing the Contribution of a Group of Predictors

Tabachnick and Fidell's (1996, pp. 174-193) regression example has number of visits to health professionals being predicted by a measure of mental health ("menheal"), physical health, and stress level. Recall that they use the log of timedrs + 1 (logtime) and physical health (logph) and the square root of stress (sqrtstr). Suppose we wish to determine the contribution that "sqrtstr" and "logph" make to predicting "logtime" when "menheal" already has been taken into account.

We have Set A = {"menheal"} and Set B = {"sqrtstr," "logph"}, so $y = 1$, $u = 2$, and $v = N - u - y - 1 = 465 - 4 = 461$. We find that $R^2_{Y.A} = .1261$ and $R^2_{Y.A,B} = .3768$, so the squared semi-partial correlation (or R^2-change) is $R^2_{Y.A,B} - R^2_{Y.A} = .2507$. The F test for this effect turns out to be $F_{2,461} = 92.7074$. This is a large, statistically significant F, and the R^2-change is large by Cohen's (1988) standards for effect sizes, but what about the precision of our estimates?

We will construct the two-sided 90% confidence interval for the corresponding partial R^2. First, the sample f^2 is obtained via rearranging Formula 4.8:

$$f^2 = (u/v)F_{u,v} = (2/461)(92.7074) = .4022.$$

Then Formula 5.19 provides us with the sample squared partial correlation $R^2_{YB.A} = f^2/(1 + f^2) = .2868$.

Assuming a fixed-score regression model and using a noncentral F routine, the resulting 90% confidence interval for the noncentrality parameter turns out to be [138.4276, 236.4989]. Using Formulas 4.10 and 5.19, we have

$$\text{lower } R^2_{YB.A} = \Delta_L/(\Delta_L + u + v + 1) = 138.4276/(138.4276 + 464) = .2298, \text{ and}$$

$$\text{upper } R^2_{YB.A} = \Delta_U/(\Delta_U + u + v + 1) = 236.4989/(236.4989 + 464) = .3376.$$

This is reasonably precise, thanks to the large sample size and the small number of predictors involved.

Example 5.6: Comparing Regression Models From Subsamples

Lee (1998) presents a study of gender and ethnic identity differences in interest in science among a sample of talented American high school participants in science and engineering-mathematics summer programs conducted in 1995-1996. He compares two regression models for each of nine scientific disciplines. Model 1 includes six predictors, mainly pertaining to gender and ethnicity. Model 2 adds three predictors to Model 1, all of which measure perceived discrepancies between participants' self-identity and identity of typical members of certain social categories as measured on a semantic-differential inventory. Lee's findings (summarized in Tables 4-6 on pp. 210-212) lead him to conclude that "all the regressions show clear negative effects of the

TABLE 5.3
95% Confidence Intervals for Partial Correlations

Discipline	Model 2 R^2	Model 1 R^2	u	v	R^2- change	F ratio	Partial R^2	Lower	Upper
Scientist	.11	.06	3	390	0.05	7.30	.05	.01	.10
Engineer	.13	.09	3	381	0.04	5.84	.04	.01	.08
Physicist	.23	.17	3	372	0.06	9.66	.07	.03	.12
Mathematician	.24	.19	3	387	0.05	8.49	.06	.02	.11
Physician	.18	.12	3	384	0.06	9.37	.07	.02	.12
Biologist	.16	.10	3	367	0.06	8.74	.07	.02	.12
Psychologist	.15	.11	3	382	0.04	5.99	.04	.01	.09
Chemist	.11	.03	3	379	0.08	11.36	.08	.03	.13
Geologist	.10	.03	3	355	0.07	9.20	.07	.02	.12

self-versus-discipline discrepancies on interest in disciplines" (p. 212). This conclusion stems partly from comparing the multiple R^2 values for the two models via the usual F tests for R^2-change. Although Lee does not report these models, he says that $p < .001$ for all the R^2-change tests.

Having found an effect for all nine disciplines, however, it would seem worthwhile to investigate whether the effect sizes are consistent or discrepant across disciplines and how precise the estimates of these effects are. Table 5.3 shows Lee's Model 1 and Model 2 multiple R^2 values, along with the relevant degrees of freedom computed from the sample sizes reported in Lee's Tables 4-6. The multiple R^2 values vary considerably, by a factor of more than 6 for Model 1 and about 2.4 for Model 2. Several of the confidence intervals (not reported here) have little or no overlap with each other. The squared semi-partial and partial correlations, however, show a rather more consistent picture regarding the effect of adding the three predictors to Model 1. Formulas 4.8, 5.17, and 5.18 have been used to complete this table, and again a noncentral F routine has been used to obtain lower and upper limits on the partial R^2 coefficients.

These intervals show considerable overlap with one another, suggesting that the effect of the three self-versus-other identity discrepancies is quite similar for all nine disciplines. Nor is the overlap the result of uniformly or unequally wide intervals, because the sample sizes are substantial and similar in all the disciplines. This is an example showing how the comparison of effects even within one study can be informed and enhanced by using confidence intervals.

Effect-Size Statistics for MANOVA and Setwise Regression

The methods in the previous section readily generalize to multivariate analysis of variance (MANOVA), multivariate (or setwise) regression, canonical correlation, and related techniques for investigating the linear association between a linear combination of independent variables and another linear combination of dependent variables. This subsection briefly outlines the necessary extensions. Wilks's lambda (Λ) statistic will be used as the centerpiece, partly because of its popularity and its ease of interpretation.

Wilks's lambda can be characterized as a ratio of determinants of two cross-product matrices, S_{effect} and S_{error}. S_{effect} is the cross-products matrix involving the effect being assessed (i.e., the effect of some linear combination of two or more independent variables), and S_{error} is the cross-products matrix of the remaining generalized variation among the dependent variables. Given this characterization, lambda is defined as

$$\Lambda = |S_{error}|/[|S_{effect}| + |S_{error}|]. \qquad [5.20]$$

A useful alternative characterization of Λ is presented in Cohen (1988, pp. 468-470), among other places, namely as the ratio of determinants of correlation matrices. Given a set X of k_X independent variables and a set Y of k_Y dependent variables, let R_X be the matrix of correlations among the variables in set X, R_Y the matrix of correlations among the variables in set Y, and R_{XY} the matrix of correlations among the variables in both sets. Then it can be shown that

$$\Lambda = 1 - |R_{XY}|/[|R_X||R_Y|]. \qquad [5.21]$$

The sample estimate of Λ will be denoted by L.

In techniques such as MANOVA, discriminant analysis, and multivariate regression, the usual significance test for Λ is based on an approximate F statistic, which can be defined just as in Formula 4.8:

$$F_{u,v} = f^2(v/u), \qquad [5.22]$$

where $f^2 = L^{-1/s} - 1$,

$$s = \sqrt{\frac{k_Y^2 k_X^2 - 4}{k_Y^2 + k_X^2 - 5}}$$

$u = k_X k_Y$ (recall that k_X and k_Y are the numbers of variables in set **X** and **Y** respectively), and

$$v = s[df_{error} - (k_Y - k_X + 1)/2] + 1 - u/2.$$

The form of the df_{error} term depends on whether variables are being partialled out from sets **X** and **Y**. If none are being partialled out, then $df_{error} = N - k_X - 1$. If a set A of k_A variables is being partialled out from sets **X** and **Y**, then $df_{error} = N - k_A - k_X - 1$. If a set A of k_A variables is being partialled out from set **X** and a set C of k_C variables is being partialled out from set **Y**, then $df_{error} = N - \max(k_A, k_C) - k_X - 1$.

Because f^2 is a generalization of the signal-to-noise ratio as used in regression and ANOVA, Formula 4.9 can be used to estimate the non-centrality parameter from the sample F statistic. A popular measure of association in MANOVA and related techniques is a simple function of L and therefore a familiar function of f^2:

$$\eta^2 = 1 - L^{1/s} = f^2/(1 + f^2) = \Delta/(\Delta + u + v + 1), \qquad [5.23]$$

as in Formula 5.6. So we can obtain confidence intervals for η^2 using Formula 4.10, just as we did in the fixed-effects ANOVA model.

Example 5.7: History and Community Attachment

Here we will use a simple data set presented by Coakes and Steed (1996, p. 142) to illustrate a MANOVA procedure. A researcher interested in determining the influence of history on variables related to community attachment surveyed community residents located across two areas of differing settlement period. Respondents were asked to complete a social support scale and a sense of community scale, and to state the size of their community networks. It was hypothesized that individuals living in the area settled prior to 1850 would have larger networks and higher levels of social support and sense of community than those residing in the area settled after 1850.

The independent variable is period of settlement, with two levels (1800-1850 and 1851-1900), and the dependent variables are sense of community, quality of social support, and network size. The usable sample is $N = 377$. Therefore $k_X = 1$ and $k_Y = 3$, $s = 1$, $u = 3$, $df_{error} = N - 2 = 375$, and $v = s[df_{error} - (k_Y - k_X + 1)/2] + 1 - u/2 = 373$. The resultant Wilks's lambda statistic has the value $L = .9486$, and $F_{3,373} = 6.7387$ ($p < .001$).

As is often the case with moderately large samples, the impressive significance level is not so impressive when we assess the effect size. We may either use Formula 5.22 to get $f^2 = L^{-1/s} - 1 = 0.0542$, or Formula 5.23 to get $\eta^2 = 1 - L^{1/s} = .0514$. The resulting 90% confidence interval for the noncentrality parameter turns out to be [6.3899, 35.6085]. Using Formula 5.23 again, we have

$$\text{lower } \eta^2 = \Delta_L/(\Delta_L + u + v + 1) = 6.3899/(6.3899 + 3 + 373 + 1)$$
$$= .0167, \text{ and}$$

$$\text{upper } \eta^2 = \Delta_U/(\Delta_U + u + v + 1) = 35.6085/(35.6085 + 3 + 373 + 1)$$
$$= .0863.$$

This interval is reasonably precise because of the relatively large sample size and the fact that there is only one predictor.

Confidence Interval for a Regression Coefficient

The methods outlined in the subsection on partial correlation are related to the usual t test and confidence interval for a regression coefficient. Here, we will develop the confidence interval for an unstandardized regression coefficient and also elucidate that relationship. When $u = 1$ (i.e., when set B contains only one variable), the t statistic that tests whether a regression coefficient differs from zero is defined by

$$t_v = b/s_{errb}, \qquad [5.24]$$

where b is the unstandardized regression coefficient and s_{errb} is its standard error. The confidence interval for b is $[b - w, b + w]$ where w is as defined in Formula 3.2, that is, $w = (t_{\alpha/2})(s_{errb})$.

Now, when set A contains y predictors, the t statistic in Formula 5.24 is related to the equivalent F statistic in the usual way:

$$t_v^2 = F_{1,v} = vf^2, \cdot$$

where $v = N - y - 1$.

From this formula and formula 4.9 we get

$$\Delta = t_v^2 = vf^2. \qquad [5.25]$$

As Formula 5.25 suggests, the noncentral $F_{1, v, \Delta}$ distribution corresponds to a noncentral $t_{v, \sqrt{\Delta}}$ distribution. Of course, the correct noncentrality parameter for the t distribution could be either a positive or a negative $\sqrt{\Delta}$.

Because the noncentral t distribution can be used to obtain confidence intervals for the standardized effect-size measure Cohen's d in any situation where a t test is legitimate, including testing individual regression coefficients, Formula 5.25 implies that we can put a Cohen's d–like measure of effect size to good use in regression. Denoting this measure by d_j for the jth regression term, it can be defined as

$$d_j = t_{jv}/\sqrt{v} = f.$$ [5.26]

So d_j is a simple function of the partial correlation coefficient. Confidence intervals for d_j or d_j^2 therefore can be constructed using the noncentral t or F distribution as appropriate.

The confidence interval for a partial correlation coefficient when $u = 1$ could be approximated using Fisher's z transformation and the normal distribution, as introduced in Chapter 3. For small samples, this approximation produces upwardly biased intervals that are also wider than those produced using the noncentral t or F approach. For larger samples, of course, this difference decreases.

Darlington (1990, pp. 208-213) makes a case for not squaring correlations when assessing their relative importance (i.e., he wants us to stick to d-units rather than d^2-units). Rosenthal (1990) argues trenchantly for not squaring correlation coefficients, on grounds that it may make important effects seem trivially small. These are arguments about how to gauge effect sizes, and this is not the place to review them. I am agnostic on this point regarding concerns that pertain to confidence intervals and their interpretation. On one hand, when the confidence interval for d_j includes 0, it is more informative (and honest) to display both the negative and positive limits instead of merely observing that the lower limit for d_j^2 is 0. On the other hand, in hierarchical regression or any situation in which we want to assess the independent contribution of a set of variables, we must use an F statistic, which in turn gives us the equivalent of d_j^2, which is a function of the squared partial multiple correlation. We can take square roots retrospectively, of course, but the basis for this decision is one's views about effect-size scaling.

Example 5.8: Assessing the Contribution of One Predictor

Mirowsky and Ross (1995) studied alternative possible explanations for apparent gender differences in self-reported distress. Part of their investigations involved two regression models in which, after including five

TABLE 5.4
Effects of Emotion𝔞l Expression and Reserve

	Sadness				Happiness			
	b	t	R^2	R^2-change	b	t	R^2	R^2-change
5-predictor model			.053				.032	
Expression	.785	21.788	.234	.181	1.220	34.157	.388	.356

predictors, they assessed the impact of an emotional "Expression" index on 2,031 respondents' sadness and happiness ratings. In their Table 4 (1995, p. 463), they reported the metric regression coefficients, their t statistics, and multiple R^2 before and after adding each predictor. The t statistics were evaluated for significance in the usual manner, each receiving three stars to indicate that $p < .001$.

The impact of Mirowsky and Ross's otherwise commendable analysis is somewhat hampered by the lack of explicit attention to effect size and confidence intervals. With such a large sample, the power to detect even small effects is high. For both t statistics shown in Table 5.4, $p < .001$, so the significance indicator is virtually uninformative about the differences in effect size for Expression and Reserve.

For Expression in the Sadness model, the partial R^2 is $.181/(1 - .053) = .1911$, so $d_{es}^2 = f^2 = 0.2363$ and $d_{es} = 0.4861$ (the subscripts here denote Expression and Sadness). The partial R^2 for Expression in the Happiness model is $.356/(1 - .032) = .3678$, so $d_{eh}^2 = f^2 = 0.5817$ and $d_{eh} = 0.7627$. This is a considerably larger effect size, but are they likely to be very different?

Another payoff for the d_j approach is in being able to assess the precision of these effect-size estimates via confidence intervals. From Formula 5.25, we can infer that for the Sadness model, $v = 2,009$, and for the Happiness model, $v = 2,006$. The 95% confidence interval for the noncentrality parameter for Expression in the Sadness model is $[19.7131, 23.8571]$, and dividing by $\sqrt{2,009}$ gives $[.4398, .5323]$, the interval for d_{es}. The 95% confidence interval for the noncentrality parameter for Expression in the Happiness model is $[31.9264, 36.3790]$, and dividing by the square root of 2,006 gives an interval for d_{eh} comprising $[.7128, .8122]$. Thanks to the large sample, these intervals are relatively narrow and are far from overlapping. The effect for Sadness is "medium," and the effect for Happiness is "large" by Cohen's (1988) standards.

Example 5.9: Evaluating a Predictor's
Contribution Across Subsamples

We return to Lee's (1998) study of influences on interest in science. One of his central findings concerns the contribution of the self-versus-discipline identity-discrepancy variable to Model 2, and as mentioned earlier, he claims that all the regressions show clear negative effects of the self-versus-discipline discrepancies on interest. This is actually not true for all the nine disciplines, however. Scientist and Physician appear to be exceptions. Could these simply reflect variability among the subsamples that is due to sampling error? One way of investigating this possibility is to use confidence intervals to indicate the range of plausible effect sizes involved.

We will undertake that evaluation here in terms of d_j^2 (i.e., f^2) units, comparing the contribution of the self-versus-discipline discrepancy (SDID) variable to that of all three identity-discrepancy variables. Lee's Tables 4-6 contain the required unstandardized regression coefficients, their standard errors, and sample sizes, which are reproduced in Table 5.5. The t statistics and f^2 values for the SDID variable are derived from this information and presented alongside the f^2 values for the three-variable contribution to Model 2 (the latter are simply converted from the partial R^2 values in Table 5.3). The rightmost two columns of the table display the lower and upper limits of 95% confidence intervals for the SDID f^2 coefficients. These intervals suggest that although the Physician (Medicine) subsample result may well be consistent with the other disciplines on this matter, the Scientist subsample SDID f^2 result does seem aberrant, even when imprecision in the estimate is taken into account.

Goodness-of-Fit Indices in Structural Equation Models

Structural equation modeling is a relatively modern technique, but until the early 1980s, models were evaluated solely in terms of a chi-square goodness-of-fit test. Because many data-sets for which structural equation modeling exercises are appropriate have large sample sizes, even models with excellent fit could be statistically rejected because of a significant chi-square test. Rather than take on the burgeoning literature on this topic, I will limit the discussion of fit indices for structural equation models to an approach due mainly to Steiger (1990), who proposed indices of fit that could be provided confidence intervals using the noncentral chi-square distribution (see also Steiger, Shapiro, & Browne, 1985).

These indices begin with a "discrepancy function," F, which is a measure of the difference between the data and a model's predictions. F is the function that is minimized by algorithms that fit structural equation models. Given a sample size of N, k groups or factors, r degrees of freedom for a particular model, and $q = N - k$, under certain conditions qF has a

TABLE 5.5

Self-Versus-Discipline-Discrepancy Contributions

Discipline	b	Standard Error	t	u	v	Three-Variable f^2	SDID f^2	Lower	Upper
Scient.	.013	.047	0.28	1	390	0.05	0.00	.00	.02
Engr.	-.204	.055	-3.71	1	381	0.04	0.04	.01	.09
Physics	-.220	.044	-5.00	1	372	0.08	0.07	.03	.13
Maths	-.267	.047	-5.68	1	387	0.06	0.08	.03	.15
Medic.	-.156	.058	-2.69	1	384	0.08	0.02	.00	.06
Biol.	-.270	.061	-4.43	1	367	0.08	0.05	.01	.11
Psych.	-.171	.048	-3.56	1	382	0.04	0.03	.01	.07
Chem.	-.185	.048	-3.85	1	379	0.09	0.04	.01	.09
Geol.	-.176	.040	-4.40	1	355	0.08	0.05	.01	.11

noncentral chi-square distribution with d degrees of freedom and non-centrality parameter $\Delta = qF$. Thus, given a confidence interval $[\Delta_L, \Delta_U]$, we immediately have a corresponding interval for F, namely $[\Delta_L/q, \Delta_U/q]$. From the earlier material on noncentral chi-square, we know to estimate the noncentrality parameter Δ by $\chi_r^2 - r$.

However, F has no penalty for model complexity, and various other indices have been proposed that do take into account complexity or parsimony. Steiger and Lind (1980) propose the RMSEA (root mean square error of approximation), defined by $(F/r)^{1/2}$. Confidence intervals for this and other similar indices are simple functions of $[\Delta_L, \Delta_U]$.

Example 5.10: Confirmatory Factor Analysis of WISC

This example is taken from Tabachnick and Fidell (1996, pp. 772-783). They apply a confirmatory factor analysis to assess the relationship between indicators of IQ and two potential underlying constructs representing IQ. There are 175 participants and 11 observed variables. They first test the independence model and find $\chi_{55}^2 = 516.237$ ($p < .01$). They then test the two-factor model they wish to confirm and find $\chi_{43}^2 = 70.236$ ($p = .005$).

The problem is that the result indicates a significant deviation of the data from the model's predictions. The goodness-of-fit indicators, however, suggest otherwise. To start with, the sample estimate of Δ is $\chi_r^2 - r = 70.236 - 43 = 27.236$. The 90% confidence interval for Δ is [8.129, 54.241]. We also have $q = N = 175$, so the sample $F = \Delta/q = 0.156$, the sample RMSEA = $(F/r)^{1/2} = 0.0602$, and the corresponding confidence interval for RMSEA is $[(\Delta_L/qr)^{1/2}, \Delta_U/qr)^{1/2}] = [0.0329, 0.0849]$. This is a fairly low RMSEA value, and the confidence interval is relatively narrow, so in conjunction with other fit indices such as GFI = .931, we may conclude that, contrary to the chi-square test result, the model fits the data quite well.

6. APPLICATIONS IN CATEGORICAL DATA ANALYSIS

In this chapter, we recapitulate and expand the material introduced in the section in Chapter 3 on the odds ratio and Example 4.4 (the noncentral chi-square distribution). First, we revisit the odds ratio and also discuss confidence intervals for differences between proportions and relative risk. Then, the RMSSE approach to developing and interpreting chi-square tests is demonstrated for one- and two-variable setups. The final section of the chapter extends this approach to assessing model comparisons in log-linear and logistic regression models.

Odds Ratio, Difference Between Proportions, and Relative Risk

The odds ratio, difference between proportions, and relative risk all are popular ways of comparing conditional distributions of the kind found in two-way contingency tables (especially 2×2 tables). Rather than compare their merits and drawbacks at length, we will merely cover methods for constructing confidence intervals and an illustrative example.

In Chapter 3, the population odds ratio was denoted by Ω and its sample estimate by W, with formula 3.14 expressing it in terms of the contingency table frequencies N_{ij} and Formula 3.15 providing an approximate standard error. The resultant confidence interval for $\ln(\Omega)$ has the form $[\ln(W) - w, \ln(W) + w]$, where $w = (z_{\alpha/2})(s_{err})$ and the standard error of $\ln(W)$ is given in Formula 3.15. As mentioned earlier, replacing the N_{ij} with $N_{ij} + 0.5$ in Formulas 3.14 and 3.15 results in better-behaved estimators, particularly if any of the N_{ij} is small. Extreme values for the odds ratio and extreme limits on the confidence interval are more drastically affected by this substitution than are values closer to an odds ratio of 1. Alternatively, exact confidence intervals are available through methods described in Agresti (1990, pp. 66-67). Because odds ratios are directly connected with parameterizations of log-linear models, confidence intervals for odds ratios are linked to confidence intervals for those parameters.

The relative risk (whose population parameter will be denoted by υ) has a sample estimate

$$RR = P_{1|1}/P_{1|2} = (N_{11}/N_{1+})/(N_{21}/N_{2+}).$$ [6.1]

The normal approximation to the sampling distribution for $\ln(RR)$ has the standard error

$$s_{err} = \sqrt{\frac{1 - P_{1|1}}{P_{1|1}N_{1+}} + \frac{1 - P_{1|2}}{P_{1|2}N_{2+}}}$$ [6.2]

As with the normal approximation method for the odds ratio, replacing the N_{ij} with $N_{ij} + 0.5$ in Formulas 6.1 and 6.2 results in better behaved estimators, particularly if any of the N_{ij} is small. For moderately large N_{ij}, an approximate confidence interval for $\ln(\upsilon)$ has the form $[\ln(RR) - w, \ln(RR) + w]$, where $w = (z_{\alpha/2})(s_{err})$.

The difference between proportions, $D = P_{1|1} - P_{1|2}$, has a large-sample approximation based on the normal approximation for the binomial distribution. The normal approximation standard error for D is

$$s_{err} = \sqrt{\frac{P_{1|1}(1 - P_{1|1})}{N_{1+}} + \frac{P_{1|2}(1 - P_{1|2})}{N_{2+}}}. \qquad [6.3]$$

Methods for small-sample and exact confidence intervals for D and RR are referred to in Agresti (1990, pp. 59-67 and p. 70, note 3.7) and reviewed at length in Santner and Snell (1980). We will deal only with the large-sample approximations here.

Example 6.1: Methadone Trials

Let us begin with a moderately large-sample study of 240 heroin addicts in Bangkok participating in a controlled trial of methadone maintenance. (I am indebted to Dr. Jeff Ward at the Australian National University for alerting me to the studies used in this example.) Vanichseni et al. (1991) compared 120 addicts randomly assigned to methadone maintenance with 120 addicts assigned to a 45-day methadone detoxification program. Among other criteria for treatment effectiveness, opiate-positive urine tests were counted and compared for the two conditions. Table 6.1 is adapted from Table 3 in Vanichseni et al. (1991, p. 1316) and divides the participants into those whose tests were opiate-positive 100% of the time versus those whose tests were not positive 100% of the time. A chi-square test of independence for this table yields $\chi_1^2 = 15.52$, $p = .00008$.

The following results were produced by adding 1/2 to each cell frequency. Starting with the odds ratio,

$$W = 2.8623, \text{ so } \ln(W) = 1.0516,$$
$$s_{err} = 0.2716,$$

and, using $z_{.025} = 1.960$, we get

$$w = 0.5323.$$

TABLE 6.1

Participants With 100% Versus < 100% Opiate-Positive Urine Tests

	100% Positive	< 100% Positive	Total
Methadone maintenance	34	86	120
45-day detoxification	64	56	120
Total	98	142	240

The 95% CI for $\ln(\Omega)$ is therefore [0.5193, 1.5839], and for Ω it is [1.6809, 4.8738].

The relative risk results yield a similar picture:

$$RR = 1.8696, \text{ so } \ln(RR) = 0.6257.$$

$$s_{err} = 0.1672,$$

and using $z_{.025} = 1.960$, we get

$$w = 0.3277.$$

The 95% CI for $\ln(\upsilon)$ is therefore [0.2980, 0.9534], and for υ it is [1.3471, 2.5946].

Finally, the difference in proportions is $D = 0.2479$, with $s_{err} = 0.0612$. Using $z_{.025} = 1.960$, we have $w = 0.1199$.

The 95% CI for D is therefore [0.1280, 0.3678].

Despite the impressive-looking significance level for the chi-square test, these confidence intervals are not very precise. The relative risk, for example, could be anywhere from about 1.35 to 2.59 times larger for the detoxification condition than for methadone maintenance, and the proportions difference could be as low as .128 and as high as .368.

The authors report percentage comparisons and chi-square tests, but no effect-size measures. This makes it difficult to compare their results with other, similar studies. For instance, Yancovitz et al. (1991) report a trial comparing methadone maintenance with a waiting-list control group over a 1-month period. They report odds ratios and a confidence interval for a table in which participants are divided into those whose urine test was

TABLE 6.2

Participants With Opiate-Positive and -Negative Urine Tests

	Positive	*Negative*	*Total*
Methadone maintenance	22	53	75
Waiting-list controls	56	38	94
Total	78	91	169

opiate-positive after 1 month versus those whose test was not positive. Table 6.2 shows their findings. A chi-square test of independence for this table yields $\chi_1^2 = 15.35$, $p = .00009$.

This result appears very similar to the chi-square value of 15.52 for the previous study; however, odds ratios or relative risk statistics may give a somewhat different picture. In this case, the odds-ratio is $W = 3.4895$ and the 95% confidence interval for it is [1.8390, 6.6213]. This is not far off the first study's odds ratio of 2.8623 and its interval [1.6809, 4.8738], but owing to the somewhat smaller sample size in this study, the interval is wider. The studies are not so nearly similar in their results as the chi-square statistics indicate, and the comparison between them is clearer using confidence intervals than point estimates. The relative risk and proportion-difference measures lead to the same conclusion.

Chi-Square Confidence Intervals for One Variable

The chi-square statistic is sometimes used to evaluate departures in the frequency distribution for a categorical variable from expected frequencies generated by a model. In Example 4.4, we saw how $\chi^2/[N(K - 1)]$ could be considered as an average squared standardized effect, as in Cramer's V^2. We may extend this idea to define a root mean square standardized effect (RMSSE) by

$$\text{RMSSE} = \sqrt{\frac{\chi^2}{N(K - 1)}}\,. \qquad [6.4]$$

or alternatively by

$$\text{RMSSE*} = \sqrt{\frac{\Delta}{N(K-1)}} \cdot \qquad [6.5]$$

If K is the number of categories in a single categorical variable, then the RMSSE will serve quite well for a one-sample chi-square effect-size measure. The procedure for constructing a confidence interval for this measure is the same as for Cramer's V except for the determination of K. Cohen (1988) does not use this approach in power analysis. Instead, he opts for an effect-size measure that can be written as

$$\omega = (\chi^2/N)^{1/2}. \qquad [6.6]$$

We may also use a version that substitutes Δ for χ^2:

$$\omega^* = (\Delta/N)^{1/2}. \qquad [6.7]$$

Clearly, RMSSE $= \omega/(K-1)^{1/2}$, and a confidence interval for the chi-square noncentrality parameter Δ also gives us an interval for both the RMSSE and ω.

Example 6.2: Evaluating Inequality

Imagine that there are two regions, Abbott and Costello, that we would like to compare for evidence of inequality in employment chances across their respective ethnic subpopulations. We have taken regional random samples of employees (600 from Abbott and 520 from Costello). Abbott contains four ethnic subpopulations (A-D), and Costello contains three subpopulations (F-H). Thus, sample size N and degrees of freedom $K - 1$ differ for these two regions. Neither the raw χ^2 nor Cohen's ω would provide an appropriate benchmark here. We adopt an equal-opportunity benchmark to provide our expected frequencies e_k, so the e_k are based on the population proportions in each region. Table 6.3 displays the observed and expected frequencies for both regions.

In fact, each of them displays inequality for the first two subpopulations listed, and the percentage difference for each of these rows is 4.8% (e.g., $[140 - 111]/600 = 0.048$). Nonetheless, because Abbott has both the larger sample and the greater *df*, the Abbott sample $\chi^2_3 = 11.614$ ($p = .009$) is larger and "more significant" than the Costello sample $\chi^2_2 = 6.696$

TABLE 6.3
Observed and Expected Frequencies for Example 6.2

Subpopulation	Abbott o_k	e_k
A	111	140
B	179	150
C	250	250
D	60	60
Total	600	600

Subpopulation	Costello o_k	e_k
F	115	140
G	305	280
H	100	100
Total	520	520

(p = .035). These figures could mislead us into thinking that Abbott has greater inequality in employment than Costello, but the percentage differences are the same (in this special instance, they can guide us because only two of them differ from 0) and so are the RMSSE values. In Abbott,

$$\text{RMSSE} = \sqrt{\frac{11.614}{600(3)}} = 0.080,$$

and in Costello,

$$\text{RMSSE} = \sqrt{\frac{6.696}{520(2)}} = 0.080.$$

Suppose we wish to construct one-sided 95% confidence intervals for both RMSSE estimates. For Costello, we obtain a 95% lower limit of 0.2429 for Δ, which translates into an RMSSE lower limit of 0.0464. For Abbott, the lower limit for Δ is 1.6680, which translates into an RMSSE lower limit of 0.0509. Because of the larger sample size from Abbott, its lower limit is slightly higher. This tells us what the two differing χ^2 values really reflect, namely, different sample sizes rather than differing degrees of inequality.

Two-Way Contingency Tables

Example 4.4 introduced a version of the root mean square standardized effect (RMSSE) in the form of the Cramer's V statistic (Formula 4.12). There are other chi-square–based statistics for two-way contingency tables, and the confidence interval for the noncentrality parameter translates into an interval for each of them. Cohen (1988) opts for $\omega = (\chi^2/N)^{1/2}$ here as well as for the univariate case, although he does note that ω is not a familiar measure of association except in 2 × 2 tables, where it reduces to the φ coefficient. It should be noted that, whereas $\omega \leq (K - 1)^{1/2}$, $V \leq 1$ (recall that $K - 1$ in this context is the smaller of $C - 1$ and $R - 1$). A second fairly popular option is to use the contingency coefficient, which can be written as

$$C = \sqrt{\frac{\chi^2}{\chi^2 + N}}.$$ [6.8]

However, its upper limit is $[(K - 1)/K]^{1/2}$ rather than 1.

Example 6.3: Attitudes Toward Miscegenation Laws

Let us reanalyze an example from attitudinal research (based on a modified version of Table 3 in Kim, 1984), in which the dependent variable is a yes-no answer to the question "Do you think there should be laws against marriages between Negroes and Whites?" and the independent variable is region in the United States (categorized as South, Central, Northeast, and West).

The chi-square test for this table yields $\chi^2_3 = 331.686$ ($p < .001$), as one might expect given the large sample and the rather obvious association between the response and region. The sample V = $[331.686/(6592)(1)]^{1/2} = .2243$. A 99% confidence interval for the noncentrality parameter turns out to be [242.6102, 429.9933]. From Formula 4.13, we have the following 99% confidence interval for V:

$$V_L = [(\Delta_L + r)/N(K - 1)]^{1/2} = [(242.6102 + 3)/(6,592)(1)]^{1/2} = .1930 \text{ and}$$

$$V_U = [(\Delta_U + r)/N(K - 1)]^{1/2} = [(429.9933 + 3)/(6,592)(1)]^{1/2} = .2563.$$

The large sample size renders this a narrow confidence interval, and we can see that the association in RMSSE terms is not all that strong.

TABLE 6.4

Data for the Miscegenation Question

Subpopulation	No	Yes	Total
South	1,086	1,386	2,472
Central	1,262	806	2,068
Northeast	952	429	1,381
West	486	185	671
Total	3,786	2,806	6,592

Alternatively, from Formula 6.8, we have C = .2189, with the 99% confidence interval

$$C_L = [(\Delta_L + r)/(\Delta_L + r + N)]^{1/2} = .1895 \text{ and}$$

$$C_U = [(\Delta_U + r)/(\Delta_U + r + N)]^{1/2} = .2483.$$

This is a very similar picture and shows the degree of association to be moderate at best. Either way, we have a much more realistic assessment of this association than we would if our attention were limited to the large chi-square and low significance level.

Effects in Log-Linear and Logistic Regression Models

We have already exploited the fact that Pearson's chi-square statistic approximates a noncentral χ^2 distribution. The same is true for the likelihood-ratio (LR) chi-square statistic used in multiway frequency analysis (log-linear models). The LR statistic is

$$G^2 = 2N\Sigma p_i\log(p_i/\pi_i), \qquad [6.9]$$

where the p_i are the observed cell probabilities and the π_i are the model's expected cell probabilities.

The noncentrality parameter has the form of G^2 in the same way that it takes the form of the Pearson chi-square statistic (cf. Agresti, 1990, pp. 435-436). In fact, because both statistics have noncentral chi-square distributions as sample sizes become large, multivariate categorical model effects may be decomposed and CIs constructed for them because of the decomposability of G^2. G^2 is decomposable in the sense that, given a log-linear model M1 that contains a simpler model M2, we may estimate the effect(s) added by M1's additional terms via

$$G^2_{(12)} = 2N\Sigma p_i \log(\pi_{1i}/\pi_{2i}),\qquad [6.10]$$

where the bracketed subscripts 1 and 2 correspond to their respective models. Therefore, we may compute power and confidence intervals for differences between nested (hierarchical) models (e.g., Agresti, 1990, pp. 241-244). The Cohen (1988) measure of $\omega = (G^2/N)^{1/2}$ or the alternative $\omega^* = (\Delta/N)^{1/2}$ occupies the same role here as with the Pearson chi-square, namely as a measure of the deviations from a model normed against sample size. Likewise, we may generalize the RMSSE approach by norming Δ on the model df as well as N. That is, we have

$$\text{RMSSE} = [G^2/(N \cdot df)]^{1/2},\qquad [6.11]$$

or

$$\text{RMSSE}^* = [\Delta/(N \cdot df)]^{1/2}.\qquad [6.12]$$

Unfortunately, the interpretation of this RMSSE is complicated by the fact that its components are not measured in standard deviation units, but instead Shannon entropies (i.e., $p_i \log[\pi_{1i}/\pi_{2i}]$). A discussion of this metric is beyond the scope of this monograph. Nonetheless, it should be noted that we still have here a dimensionless effect-size measure whose magnitude is independent of sample size and table size.

Example 6.4: A Three-Way Table

In Table 6.5 (from Darlington, 1990, p. 464), three variables have been cross-tabulated. The data concern students in a clinical psychology program, and the row variable is whether they graduated or dropped out of the program. The students have been divided into males and females because the evaluators of the program were concerned about a higher dropout rate for males (which you can see simply by inspecting the table or by computing column percentages).

A simple *independence* model (equivalent to the chi-square test when there are just two variables) includes all possible first-order terms. In log-linear model notation, this model is

$$\ln(e_{ijk}) = \theta + \lambda_{Ai} + \lambda_{Bj} + \lambda_{Ck},$$

TABLE 6.5
Graduation Data: Observed Counts

	Counselor A		Counselor B		Counselor C	
	Male	Female	Male	Female	Male	Female
Grad.: Yes	36	27	20	37	15	33
Grad.: No	18	7	19	13	32	19

Source: From Darlington (1990), p. 464.

TABLE 6.6
Independence Model Expected Counts

	Counselor A		Counselor B		Counselor C	
	Male	Female	Male	Female	Male	Female
Grad.: Yes	27.17	26.39	27.48	26.69	30.57	29.69
Grad.: No	17.47	16.97	17.67	17.16	19.65	19.09

Source: From Darlington (1990), Table 19.3.

where A = Gender, B = Graduation, and C = Counselor. A log-linear program computes expected frequencies, taking into account the marginal totals for each of these three variables while assuming that they are all independent of each other. The expected frequencies are those shown in Darlington's Table 19.3, reproduced here as Table 6.6.

This table has the same odds of graduation in every subtable. That is, $27.17/17.47 = 26.39/16.97 = 27.48/16.67 = \cdots = 1.555$. In short, in this model the odds of graduating are the same regardless of gender or counselor. As it turns out, the LR chi-square for the independence model is $G^2 = 33.81$ with 7 df, which is significant, so there must be some kind of association between two or more of the variables. Our next step is to add an interaction term, Gender × Graduation. Our model now is

$$\ln(e_{ijk}) = \theta + \lambda_{Ai} + \lambda_{Bj} + \lambda_{Ck} + \lambda_{AiBj}.$$

The chi-square value for this model turns out to be $G^2 = 21.39$ with 6 df. This chi-square is also significant, so the model still does not fit the data well.

But is it an improvement over the first model? To answer that question, we need to assess the change in chi-square, taking the change in degrees of freedom into account. Thanks to the decomposability of G^2, we just take the difference between the chi-squares ($33.81 - 21.39 = 12.42$), which gives us another chi-square. Its degrees of freedom are equal to the difference between the dfs associated with the two models ($7 - 6 = 1$). A chi-square of $G^2_{(12)} = 12.42$ with 1 df is significant, so the new model definitely is an improvement. This procedure is almost directly analogous to the F test for the significance of the partial R^2 in hierarchical or stepwise regression, and it is a specific type of model comparison.

How large an effect does the association between Gender and Graduation contribute? Because $df = 1$ in this case,

$$\text{RMSSE*} = w* = (\Delta/N)^{1/2} = ([12.42 - 1]/276)^{1/2} = 0.2034.$$

A one-sided 95% confidence interval for the noncentrality parameter yields a lower limit of $\Delta = 3.5330$, which translates into a lower 95% limit on RMSSE and ω of 0.1131. This information would be quite useful in comparing the results of this study with a replication, or in planning a replication study.

Example 6.5: Logistic Regression

In logistic regression, there has been something of a cottage industry in summary statistics for the assessment of goodness of fit and model comparison (cf. Hosmer & Lemeshow, 1989, chap. 5). Although the practice of using large-sample asymptotic theory to construct confidence intervals for individual coefficients in logistic regression models is straightforward and commonplace, to my knowledge little has been done in this vein for model comparison or goodness-of-fit statistics. Here, we will only scratch the surface of this topic.

The $G^2_{(12)}$ statistic from Formula 6.10 is used to compare a model with a reduced model containing a subset of the first model's predictors. The RMSSE statistic in Formula 6.11 is one way of norming the gain in fit measured by G^2 with respect to sample size and degrees of freedom. Another suggestion has been an R^2-type measure (actually a proportional reduction of error measure). As before, given a logistic regression model M1 to be compared with a simpler model M2 whose predictors form a subset of those in M1, we define this measure by

$$R^2_L = 100(L_2 - L_1)/L_2, \qquad [6.13]$$

where L_2 is the log-likelihood for M2 and L_1 is the log-likelihood for M1. In fact, $L_2 - L_1 = G^2_{(12)}/2$ from Formula 6.10. If we are willing to regard the

M2 model as fixed, then a confidence interval may be obtained for R_L^2 via the noncentral chi-square method used for the RMSSE. However, the RMSSE requires no such restriction and is therefore more flexible.

One of Hosmer and Lemeshow's examples is a hypothetical study assessing the impact of gender and age on coronary heart disease (CHD) status, which is measured dichotomously. Age has four categories, and the study includes 100 subjects (Hosmer & Lemeshow, 1989, p. 48). In one version of the study (Table 3.15, p. 66), the effect of adding age to the model yields $G_{(12)}^2 = 24.54$. Suppose in another study with 450 subjects and an age variable that was given six categories instead of four, adding age to the same model yielded $G_{(12)}^2 = 52.78$. Both of these are significant effects. How can we comparatively assess the effect size and precision of estimates in these two studies, given the different sample sizes and numbers of categories for the age variable?

In the first study, $N = 100$ and $df = 4 - 1 = 3$, so from Formula 6.11 we have

$$\text{RMSSE} = [24.54/(100 \cdot 3)]^{1/2} = 0.286.$$

A 95% confidence interval for Δ is [7.4175, 45.4134], so from Formula 6.11 we find the interval for the RMSSE is [0.186, 0.402].

In the second study, $N = 450$ and $df = 6 - 1 = 5$, so from Formula 6.11 we have

$$\text{RMSSE} = [52.78/(450 \cdot 5)]^{1/2} = 0.153.$$

A 95% confidence interval for Δ is [24.7278, 80.5874], and the corresponding interval for the RMSSE is [0.115, 0.195]. The effect of age in the second study does not appear to be as strong as in the first, but owing to the larger sample size, the precision of the estimate in the second study is considerably higher.

7. SIGNIFICANCE TESTS AND POWER ANALYSIS

Significance Tests and Model Comparison

We have already seen that confidence intervals include information relevant to null hypothesis significance tests. If a 95% confidence interval for the difference between two means from independent samples contains 0, then we immediately know that we cannot reject the null hypothesis at the

.05 level. However, we also know that we cannot reject any of the nonzero hypothetical values for this difference that lie inside the interval. Any value inside the interval could be said to be a *plausible* value; those outside the interval could be called *implausible*. Smithson (2000, p. 177) formalizes this terminology; many other writers have used it more or less informally.

What advantage is there in having a display of all the hypothetical values that cannot be rejected via a significance test? One benefit is that it paves the way toward assessing competing hypotheses, including "vague" ones that eschew pointwise predictions. Returning to Example 3.1, in which a random sample of 20 adults from a population whose mean IQ is 100 is given a treatment that is claimed to increase IQ and their posttreatment mean IQ is 106.5 with a standard deviation of 15, we computed the 95% confidence interval for μ of [99.48, 113.52]. A no-effects hypothesis (i.e., $\mu = 100$) and a 5-point increase hypothesis (i.e., $\mu = 105$) are both plausible, and therefore neither can be rejected (although we would be silly to believe in either of them).

Now suppose we had two competing but somewhat vague hypotheses about the treatment's effect:

A. There will be a 5- to 15-point increase (i.e., $105 < \mu < 115$), and
B. There will be a 15- to 20-point increase (i.e., $115 < \mu < 120$).

The confidence interval tells us that we can definitely reject B because none of its predicted values are plausible. What about A? Its predicted values from 105 to 113.52 are plausible, so it cannot be ruled out, even though its predicted values between 113.52 and 115 are outside the confidence interval. Clearly there is a trade-off between the vagueness of a hypothesis and the extent to which we are able to decisively reject it or not. Nevertheless, we are in a much better position to evaluate such hypotheses with confidence intervals than with pointwise null hypothesis significance tests.

Model comparison is a generalization of significance testing, in which a more complex model is compared with a simple model whose terms form a subset of the first model. We have seen that confidence intervals may be used to good effect on the summary statistics that are used to assess gains in goodness of fit or predictivity that are achieved with more complex models, both for the purpose of establishing how precisely those gains have been estimated and for comparisons with gains from other models in the same study or the same model in different studies. Particularly in cases such as multiple partial R^2 and the chi-square change statistics for logistic regression and log-linear models, the practice of constructing confidence intervals for the effect of groups of variables (versus constructing them for only one variable at a time) can only be beneficial.

Power and Precision

The power of a statistical test of a null hypothesis is the probability of correctly rejecting that null hypothesis, for a specific effect size and Type I error (significance) criterion. Ideally, one is supposed to perform a power analysis prior to collecting data, usually to ascertain the smallest effect that can be detected in a study with a specified sample size, significance criterion, and power criterion. Many researchers, however, support the practice of post hoc power analysis, especially in the face of an unexpected failure to reject a null hypothesis. We will deal with both kinds of power analysis here in relation to confidence interval estimation.

Some authors (e.g., Oakes, 1986, p. 68) have claimed that power is inextricably bound up with significance tests, and that if confidence intervals replaced significance tests, there would be no need for power calculations. Others, such as Steiger and Fouladi (1997), advocate combining power analysis with interval estimation, including confidence intervals for power. I will take an agnostic position on this matter, as I have on banning significance tests. At the very least, researchers and students reading the older significance-oriented literature will need to understand the relationship between confidence interval width and power. At best, considerations of both power and precision of estimates may help researchers design and evaluate studies.

Briefly, power analysis involves the same sampling distributions as confidence intervals do. For example, any time power analysis on standardized effects can be done, so can noncentral confidence interval estimation. This fact makes it all the more remarkable that software tools for noncentral confidence interval estimation were not made widely available at the same time as tools for power analysis, and perhaps is yet another testament to the seductiveness of the Neyman-Pearson-Fisherian significance testing framework.

Nonetheless, confidence intervals provide different information from power analysis; as we shall see, high power does not always entail precise estimation, nor vice versa. The key to understanding the difference between them is the realization that although confidence (or significance) level and sample size affect both power and interval width, effect size affects only power.

For instance, suppose we have a repeated-measures study with a small sample of $N = 9$ subjects measured on two occasions, and we examine how power and interval width vary as a function of Cohen's d. Table 7.1 displays

TABLE 7.1
Effect Size, Noncentral CI, and Power (two-tailed $\alpha = .05$)

t	df	d	Power	Lower d	Upper d	Precision
1.9	8	0.633	.3872	−0.1036	1.3386	1.4422
2.1	8	0.700	.4553	−0.0520	1.4184	1.4703
2.3	8	0.767	.5246	−0.0013	1.4995	1.5008
2.5	8	0.833	.5931	0.0480	1.5817	1.5337
2.7	8	0.900	.6587	0.0967	1.6651	1.5684
2.9	8	0.967	.7197	0.1449	1.7495	1.6046
3.1	8	1.033	.7748	0.1924	1.8348	1.6425
3.3	8	1.100	.8230	0.2390	1.9211	1.6821
3.5	8	1.167	.8641	0.2851	2.0081	1.7230

the results from $d = 0.633$ to $d = 1.167$. Power increases with effect size, of course. But the precision of our estimate does not increase, as indicated in the rightmost column of Table 7.1; in fact, there is a mild trend in the opposite direction. This illustrates the need to use confidence intervals along with power, both in designing a study and in assessing findings. A study that has a small sample size might end up with apparently high power given a large effect, but a confidence interval will reveal that the estimation of the effect size still lacks precision, and therefore, so does any post hoc estimate of power.

Now suppose we have reason to expect to find an effect size of about $d = 0.5$. Given a 95% confidence level, we would need a sample size of at least 44 to detect this effect with power greater than .9. But for $N = 44$, a 95% confidence interval for d turns out to be [0.184, 0.811]. Given that Cohen's (1988) benchmarks characterize $d = 0.2$ as "small" and $d = 0.8$ as "large," this confidence interval tells us that a sample size of 44 would enable us to say only that the effect size was anywhere from "small" to "large."

If we require instead a 95% interval whose width is 0.4 (i.e., a bit more than the distance between two adjacent Cohen's benchmarks), then for $d = 0.5$, the minimum sample size requirement is 108. The resulting interval for d is [0.299, 0.699]. If we consider higher values of d, then the difference between sample sizes required for respectable power versus precision becomes greater.

If we consider small effect sizes, however, the samples required for respectable power become larger than those needed for reasonable precision. For $d = 0.2$, for instance, the sample size needed for a 95% interval whose width is 0.4 is 100, but power is only .508. For power to be increased to .9, we require a sample of at least 265. The 95% confidence interval for d given this sample size is [0.078, 0.321].

Overall, however, what would appear to be modest criteria for precision require larger sample sizes than conventional criteria for power. This may seem discouraging insofar as narrow confidence intervals for even medium effect sizes may require what for many researchers are unattainably large samples. Schmidt (1996) made a similar point about power requirements, and his remarks also apply here. Reliance on significance testing has imbued many researchers with the false assumption that every study should and can support a conclusion (e.g., "Does the treatment have an effect?"). A more reasonable and sober assessment is that a collection of studies, each of them properly conducted, may reach such a definite conclusion provided that their results are combined via careful meta-analysis. Confidence intervals contribute more to the accumulation of scientific knowledge than significance tests because they avoid the misleading "vote-counting" to which significance testing is prone.

Designing Studies Using Power
Analysis and Confidence Intervals

It should be clear from the examples and arguments thus far that researchers would do well to consider both power and precision requirements in designing their studies. As far as I am aware, this is very rarely done. Such a practice, however, would improve understanding of both the limits of individual studies and the role of meta-analysis. It would require establishing viable benchmarks for confidence interval width, some of which undoubtedly will need to differ among subdisciplines. My own preference is to explore power before conducting a study and to link that exploration with an investigation of confidence interval width as well. After all, reasonable criteria for deciding on sample size should include the precision of sample estimates as well as the power to reject a null hypothesis. I shall denote this kind of assessment by the phrase "precision analysis."

Example 7.1: Partial Correlation

Consider Cohen's example (1988, pp. 435-436) in which, after entering eight covariates into a regression model, two predictors are to be entered and are expected to have $R^2_{YB.A} = .1$. When $N = 148$ and $\alpha = .01$, he finds the power of the study to detect this effect is .84. Given the somewhat stringent α level, this appears to be a respectable amount of statistical power.

Now suppose we actually observed a *sample* $R^2_{YB.A} = .1$. How precise would this estimate of the contribution from the two predictors be? With $u = 2$ and $v = 148 - 2 - 8 - 1 = 137$, a 98% two-sided confidence interval for partial R^2 turns out to be [.0115, .2146]. Cohen's (1988) benchmarks for multiple R^2 are .01 = small, .09 = medium, and .25 = large. Our confidence interval nearly spans the range from "small" to "large."

The major difference between the questions addressed by power analysis and precision analysis is that with precision analysis, we start with positing an *observed* effect size rather than assuming a population value for the effect size. In this sense, as well as in their primary objectives, precision analysis and power analysis are complementary.

When the goal is to explore the impact of sample size on confidence interval width, I advocate using an unbiased measure of effect size even if the confidence interval is not based on that estimator. For example, it is better to use the adjusted R_a^2 rather than R^2 for the observed effect size because the less biased R_a^2 better approximates the goal of keeping the true effect size constant across different sample sizes and numbers of predictors.

The researcher must also be careful about the impact of effect size on confidence interval width versus its impact on power, especially for effect-size measures that are bounded below and above. For a given sample size, for instance, the widest confidence interval for multiple R^2 seldom occurs for the lowest plausible value. Consider a four-predictor model with a sample size of $N = 155$. A 95% confidence interval for R^2 is widest in the neighborhood of .3, as Figure 7.1 shows. Power, on the other hand, simply increases with R^2. In designing a study in which the researcher expected R^2 to be anywhere from .1 to .5, power considerations would require focusing on $R^2 = .1$, whereas precision considerations would require exploring inter-mediate values near $R^2 = .3$.

Example 7.2: Designing a Regression Study With Multiple Predictors

This example is not a fully fledged attempt at determining sample size but does illustrate the issues that would be considered in such a design. To begin with, let us assume that one of the main purposes of our study is prediction using a linear regression approach. We would need to ask how many predictors could end up in our final regression model. Suppose we believe the number of predictors involved in a regression model could vary from 3 to 10. Both power and precision considerations would motivate us to focus on the case in which $u = 10$, because that is the worst-case condition on both counts.

Next, we would need to think about the plausible range of effect sizes. Suppose we have reason to think that R^2 will probably turn out to be some-where between .1 and .5. In the absence of any other guidelines, we could start by exploring sample-size requirements in this range. Power consider-ations would motivate a focus exclusively in the vicinity of $R^2 = .1$, but pre-cision would require investigating the larger values of R^2 in the range.

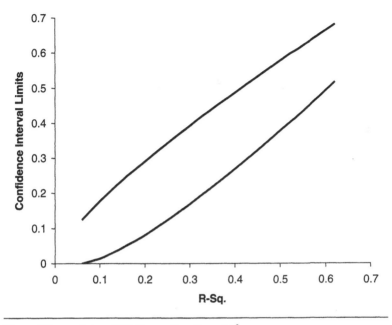

Figure 7.1   Interval Width as a Function of R^2

We also would need to decide on a confidence level (say, .95) and a criterion for power (say, .90). Power curves when $u = 10$, confidence level is .95, and R^2 is in the neighborhood of .1 are shown in Figure 7.2. For the sake of convenience, a fixed-score regression model has been used here, and therefore these curves are derived from the appropriate noncentral F distributions. The minimum sample size is clearly quite sensitive to small variations in R^2. When $R^2 = .08$, the minimum sample size is about 245, whereas for $R^2 = .1$, that drops to about 195, and for $R^2 = .12$, it declines to around 160. Note that for a fully fledged investigation into sample size requirements, we would also want to investigate how sensitive power is to confidence levels in the region around .95.

Taking precision into account entails deciding on a criterion for acceptable confidence interval width. Because this criterion has received little serious attention in the social science literature, we shall have to adopt a fairly arbitrary example of such a criterion for illustrative purposes. Recalling that Cohen's (1988) benchmarks are that $R^2 = .01$ is "small," $R^2 = .09$ is "medium," and $R^2 = .25$ is "large," we might require that when R^2 is around .25, the lower limit of the 95% confidence interval should exceed .09, which is roughly akin to being able to distinguish a "large"

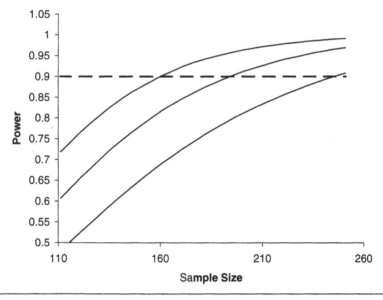

Figure 7.2   Power Curves for R^2 = .08, .1, and .12

from a "medium" effect. In the same vein, when R^2 is around .1, the upper limit of the 95% confidence interval should fall below .25 and the lower limit should exceed .01.

It turns out that the second criterion is the more severe of these two, but even it is not as strict as the power requirement we imposed earlier. If R_a^2 = .1, then a sample size of only 118 is sufficient to result in a lower confidence interval limit that exceeds .01, and the requisite sample size does not rise to 160 until R_a^2 is reduced to .08. This outcome merely reflects the fact that our precision criteria are quite modest.

If we take our power requirement seriously and conclude that sample sizes in the vicinity of 190-240 would be reasonably likely to ensure power of .9 or more, it would be useful to ascertain how the confidence interval width behaves for such samples for R_a^2 from .1 to .5. Figure 7.3 shows the resulting interval widths (precision) as a function of sample size for three values of R_a^2 (.1, .3, and .5). These curves indicate that precision is approximately equally variable across the sample size and R^2 ranges. Precision is nearly linear in sample size, but of course it is quite nonlinear as a function of R_a^2. The widest intervals occur when R_a^2 = .3, and an in-depth exploration of the trade-offs between sample size and precision would necessitate concentrating on values near .3.

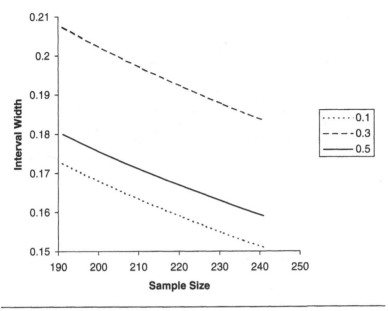

Figure 7.3   Precision (Interval Width) Curves for R^2 = .1, .3, and .5.

Some writers also recommend examining how precision varies with confidence level (e.g., Cox & Hinkley, 1974), and this certainly seems advisable. It is easy to see that the increase in interval width accelerates as confidence level increases. In our example, when R_a^2 = .3, decreasing the confidence level by .03 from .95 to .92 narrows confidence intervals by about .02 for samples of around 200, but increasing confidence by .03 from .95 to .98 widens the intervals by about .035. A similar point holds regarding the effect on power of increasing or decreasing confidence levels. Pragmatic researchers might willingly trade a small decrease in confidence level to achieve precision or power criteria, or invest in a larger sample to increase confidence level while maintaining a given level of precision or power. In any event, the information gained by investigating such trade-offs before actually conducting a study can only benefit the researcher.

Although the examples in this section have emphasized partial and multiple correlation, that should not be taken to mean that these are the best or most appropriate quantities to focus on in designing studies. Each study should be designed taking into account estimates and CIs for those parameters that are meaningful and interesting for the purposes of that study.

Confidence Intervals for Power

Finally, as mentioned earlier in this chapter, confidence intervals for a statistic also can be transformed into confidence intervals for power. The place for this is in post hoc analyses. A typical application would be when a study fails to yield an expected significant finding and the possibility arises that this may be due to insufficient statistical power. Likewise, this stratagem would be informative in evaluating published studies. The main idea is simply to find the power of detecting effects of the sizes indicated by the limits of the confidence interval for them, given the sample size and significance criterion (or confidence level). One example should suffice to illustrate the principles involved.

Example 7.3: Simple Correlation

Suppose a researcher observed a sample value for $R = .35$, with a sample size of $N = 60$. A 95% confidence interval for R would be [0.1054, 0.5546]. We could ask, after the fact, what the power is to detect an R within this confidence interval. In so doing, we would need to find the power to detect $R = 0.1054$ and $R = 0.5546$, given $N = 60$ and a significance criterion of .05.

For $R = .1054$, Formula 3.11 yields a transformed $R' = 0.1058$ and Formula 3.12 gives a standard error of $s_{R'} = .1325$. The null hypothesis of $R = 0$ is equivalent to a null hypothesis that $R' = 0$. To reject this null hypothesis using a one-tailed test, we would need to observe a sample R' of at least $(z_{.05})(s_{R'}) = (1.645)(.1325) = .2180$. For a normal distribution with a mean of 0.1058 and standard deviation of .1325, the probability of observing a value higher than .2180 is .1986. That probability is the power of our test if the population $R = .1054$ (i.e., if $R' = .1058$).

Likewise, for $R = .5546$, we get $R' = 0.6250$. For a normal distribution with a mean of 0.6250 and standard deviation of .1325, the probability of observing a value higher than .2180 is .9989. This is the power of our test if the population $R = .5546$ (i.e., if $R' = .6250$). Given an observed $R = .35$ with a sample size of $N = 60$ and significance criterion of .05, a 95% confidence interval for power is [.1986, .9989]. This interval is very wide because the sample size is small and the value of R is not near 0 or 1.

8. CONCLUDING REMARKS

For many research areas in the human sciences, confidence intervals have much to recommend them. They are effective inferential statistical summaries and can be used for both cumulative knowledge and decision making or hypothesis testing. Although use of confidence intervals is arguably superior to the traditional significance testing approach, it is not a panacea. We have already seen that confidence intervals have some weaknesses, the greatest of which probably is the temptation to interpret the long-run coverage rate as if it applies to each individual confidence interval.

Confidence intervals also raise issues that often remain unaddressed in both basic and applied research. Although this is in itself not bad, such issues may impede the adoption of confidence intervals. Many research areas and even entire disciplines face unresolved problems in establishing meaningful guidelines for confidence levels and acceptable interval widths. Matters are complicated somewhat by the fact that interval width may covary somewhat with effect size but does not always covary with power. In Chapter 7, I suggested using Cohen's effect-size benchmarks as limits for confidence interval widths in determining minimum sample sizes. Obviously, this may not be the best criterion for some purposes, but in the absence of any reasonable alternative, it could be a useful guideline. For example, a researcher designing a study might ascertain the minimum sample size required so that for a "medium" effect size, the confidence interval excludes the "small" and "large" effect-size values. The issue of acceptable interval widths will always depend, however, on the effect sizes that the researcher believes relevant. Small effects, for instance, will entail concerns about sample sizes required to exclude 0 from the interval.

Because interval estimates involve estimates of variation, they are bound to highlight difficulties in obtaining accurate variance estimates. Two frequently overlooked problems of this kind stem from measurement error and complex sample designs. Neither of these has been addressed in this monograph. Virtually all the confidence intervals developed herein do not incorporate measurement error, and they are based on simple random sampling models.

Starting with measurement error, the nub of the problem is that any variance estimate or effect-size estimate from sample data is affected not only by sampling error (i.e., "true" variation) but also by measurement error (i.e., unreliability). Unreliability inflates observed variances and attenuates observed effects of independent variables (and therefore observed effect sizes). For example, if variable X has reliability r_x and Y has reliability r_y, then the expected observed correlation between X and Y will be attenuated in the following fashion:

$$R_{xy} = \rho_{xy}\sqrt{r_x}\sqrt{r_y}, \qquad [8.1]$$

where ρ_{xy} is the true population correlation. For many research purposes, there is little harm in these effects as long as the researcher bears them in mind and makes inferences about observed rather than true-score outcomes. In situations where reliability estimates are available, however, the researcher may wish to separate sampling from measurement error. In prediction, if we are predicting Y conditional on X measured without error, then we would correct for attenuation, but if we are predicting Y conditional on X measured with error, then we would not.

Cohen (1988) and Levin and Subkoviak (1977) present contrasting views on this subject regarding effect sizes in power analysis. Because power analysis begins with assumed population effect sizes, assuming error-free measurement amounts to overestimating power and underestimating required sample sizes. Levin and Subkoviak (1977) contend that researchers should correct for this in power analysis, whereas Cohen maintains that such a correction has no real merit.

Pointwise estimates and confidence intervals, on the other hand, are based on sample statistics and therefore will lead the researcher to underestimate population effect sizes and interval widths. For a simple correlation, as Hunter and Schmidt (1990, pp. 118-122) point out, we would apply the correction for reliability implied by Formula 6.13 to the endpoints of the confidence interval. The transformed interval will always be wider than the original. The distinction here is between the population *uncorrected* correlation and the population *corrected* correlation. Researchers will need to decide to which parameter they want to refer.

Now let us turn to complex sample designs, by which I mean probabilistic samples that are multistage, hierarchical, stratified, or model-driven. These designs affect estimator bias, variability, and confidence interval coverage-rates. For instance, multistage cluster samples inflate sample variances so that standard errors based on simple random sampling models are underestimates, whereas stratified random samples decrease variances of any variables correlated with the strata. For those variables, confidence intervals based on simple random sampling models will be too narrow in cluster samples and too wide in stratified samples. Interested readers should consult the extensive literature on survey sampling and sample design effects.

Meta-analysis has been mentioned several times in this monograph. Like the confidence interval, meta-analysis is concerned with statements about the variability of a population parameter, but meta-analytic statements involve accumulations of studies. Hunter and Schmidt's (1990), Hedges and Olkin's (1985), and Rosenthal's (1991) frameworks involve cumulative estimates of effect sizes across studies, taking into account a variety of error-generating artifacts and second-order sampling error. They are certainly compatible with the confidence interval approach, and their proponents explicitly incorporate it in their methods. For some time, the primary exception to all this was the reliance on a significance test for detecting effect-size heterogeneity and moderator model misspecification. Recently, however, techniques have been developed for power analysis (Hedges & Pigott, 2001) and confidence intervals for heterogeneity statistics (Smithson, 2002).

Finally, a few remarks are in order about available software, although such remarks are bound to become obsolete quite rapidly. As of this writing, most popular statistical packages (e.g., Minitab, SPSS, SAS) provide confidence intervals based on standardizable distributions such as the normal and t distributions. It is less common for such packages to compute intervals based on nonstandardizable distributions such as the noncentral t, χ^2, and F, even though some of them include calculators for those distributions' density functions. Statistica is an exception, with an add-on package for CI and power calculations involving noncentral distributions. SPSS, SAS, SPlus, and R routines for computing confidence intervals and power for the noncentrality parameters and related statistics for the noncentral t, χ^2, and F distributions are available free of charge from the author via his Web page (www.anu.edu.au/psychology/staff/mike/Index.html), however, and Steiger and Fouladi's R2 program for multiple R^2 intervals may be obtained from Steiger's Web page (www.interchg.ubc.ca/steiger/homepage.htm).

REFERENCES

AGRESTI, A. (1990). *Categorical data analysis*. New York: Wiley.

ALTMAN, D. G., MACHIN, D., BRYANT, T. N., & GARDNER, M. J. (2000). *Statistics with confidence: Confidence intervals and statistical guidelines* (2nd ed.). London: British Medical Journal Books.

BAILAR, J. C., & MOSTELLER, F. (1988). Guidelines for statistical reporting in articles for medical journals: Amplifications and explanations. *Annals of Internal Medicine, 108,* 266-273. [Available online at www.acponline.org/journals/resource/guidelines.htm]

BALAKRISHNAN, N., & MA, C. W. (1990). A comparative study of various tests for the equality of two population variances. *Journal of Statistical Computation and Simulation, 35,* 41-89.

BURDICK, R. K., & GRAYBILL, F. A. (1992). *Confidence intervals on variance components*. New York: Marcel Dekker.

BURDICK, R. K., MAQSOOD, F., & GRAYBILL, F. A. (1986). Confidence intervals on the intraclass correlation in the unbalanced one-way classification. *Communications in Statistics—Theory, Methods, 15,* 3353-3378.

COAKES, S. J., & STEED, L. G. (1996). *SPSS for Windows: Analysis without anguish*. Brisbane: Jacaranda Wiley.

COHEN, J. (1962). The statistical power of abnormal-social psychological research: A review. *Journal of Abnormal and Social Psychology, 65,* 145-153.

COHEN, J. (1988). *Statistical power analysis for the behavioral sciences* (2nd ed.). Hillsdale, NJ: Erlbaum.

CONOVER, W. J. (1980). *Practical nonparametric statistics* (2nd ed.). New York: Wiley.

COX, D. R., & HINKLEY, D. V. (1974). *Theoretical statistics*. London: Chapman and Hall.

CUMMING, G., & FINCH, S. (2001). A primer on the understanding, use, and calculation of confidence intervals that are based on central and noncentral distributions. *Educational and Psychological Measurement, 61,* 532-574.

DARLINGTON, R. (1990). *Regression*. New York: McGraw-Hill.

DIACONIS, P., & EFRON, B. (1983). Computer-intensive methods in statistics. *Scientific American, 248*(5), 116-130.

EFRON, B., & TIBSHIRANI, R. J. (1993). *An introduction to the bootstrap*. London: Chapman and Hall.

FIDLER, F., & THOMPSON, B. (2001). Computing correct confidence intervals for ANOVA fixed- and random-effects effect sizes. *Educational and Psychological Measurement, 61,* 575-604.

FLEISHMAN, A. I. (1980). Confidence intervals for correlation ratios. *Educational and Psychological Measurement, 40,* 659-670.

GERRITY, M. S., EARP, J.A.L., & DeVELLIS, R. F. (1992). Uncertainty and professional work: Perceptions of physicians in clinical practice. *American Journal of Sociology, 97,* 1022-1051.

HARLOW, L., MULAIK, S., & STEIGER, J. H. (Eds.). (1997). *What if there were no significance tests?* Hillsdale, NJ: Erlbaum.

HARVILLE, D. A., & FENECH, A. P. (1985). Confidence intervals for a variance ratio, or for heritability, in an unbalanced mixed linear model. *Biometrics, 41,* 137-152.

HEDGES, L. V., & OLKIN, I. (1985). *Statistical methods for meta-analysis*. New York: Academic Press.

HEDGES, L. V., & PIGOTT, T. D. (2001). The power of statistical tests in meta-analysis. *Psychological Methods, 6,* 203-217.

HENKEL, R. (1976). *Tests of significance*. Sage University Papers Series on Quantitative Applications in the Social Sciences, 07-004. Thousand Oaks, CA: Sage.

HOSMER, D. W., & LEMESHOW, S. (1989). *Applied logistic regression.* New York: Wiley.

HOWELL, D. C. (1997). *Statistical methods for psychology* (4th ed.). Belmont, CA: Duxbury.

HUNTER, J. E., & SCHMIDT, F. L. (1990). *Methods of meta-analysis: Correcting error and bias in research findings.* Thousand Oaks, CA: Sage.

KEMPTHORNE, O., & FOLKS, L. (1971). *Probability, statistics, and data analysis.* Ames: Iowa State University Press.

KENDALL, M. G. (1949). Reconciliation of theories of probability. *Biometrika, 36,* 101-116.

KENDALL, M. G., & STUART, A. (1979). *The advanced theory of statistics* (Vol. 2). New York: Macmillan.

KIM, J.-O. (1984). PRU measure of association for contingency table analysis. *Sociological Methods & Research, 13,* 3-44.

LEE, J. D. (1998). Which kids can "become" scientists? Effects of gender, self-concepts, and perceptions of scientists. *Social Psychological Quarterly, 61,* 199-219.

LEHMAN, E. L. (1986). *Testing statistical hypotheses* (2nd ed.). New York: Wiley.

LEVIN, J. R., & SUBKOVIAK, M. J. (1977). Planning an experiment in the company of measurement error. *Applied Psychological Measurement, 1,* 331-338.

LINDLEY, D. V. (1965). *Introduction to probability and statistics from a Bayesian viewpoint, Part 2: Inference.* Cambridge, UK: Cambridge University Press.

LOCKHART, R. S. (1998). *Introduction to statistics and data analysis for the behavioral sciences.* New York: W. H. Freeman.

MAXWELL, S. E., & DELANEY, H. D. (1990). *Designing experiments and analyzing data: A model comparison perspective.* Belmont, CA: Wadsworth.

MEEHL, P. E. (1967). Theory testing in psychology and physics: A methodological paradox. *Philosophy of Science, 34,* 103-115.

MENDOZA, J. L., & STAFFORD, K. L. (2001). Confidence intervals, power calculation, and sample size estimation for the squared multiple correlation coefficient under the fixed and random regression models: A computer program and useful standard tables. *Educational and Psychological Measurement, 61,* 650-667.

MIROWSKY, J., & ROSS, C. E. (1995). Sex differences in distress: Real or artifact? *American Sociological Review, 60,* 449-468.

NEYMAN, J. (1935). On the problem of confidence intervals. *Annals of Mathematical Statistics, 6,* 111-116.

NEYMAN, J. (1937). Outline of a theory of statistical estimation based on the classical theory of probability. *Philosophical Transactions of the Royal Society,* Series A, *236,* 333-380.

OAKES, M. L. (1986). *Statistical inference: A commentary for the social and behavioral sciences.* New York: Wiley.

ROSENTHAL, R. (1990). How are we doing in soft psychology? *American Psychologist, 45,* 775-777.

ROSENTHAL, R. (1991). *Meta-analytic procedures for social research* (rev. ed.). Newbury Park, CA: Sage.

ROSENTHAL, R., & ROSNOW, R. L. (1985). *Contrast analysis: Focused comparisons in the analysis of variance.* Cambridge, UK: Cambridge University Press.

ROZEBOOM, W. W. (1960). The fallacy of the null hypothesis significance test. *Psychological Bulletin, 57,* 416-428.

SANTNER, T. J., & SNELL, M. K. (1980). Small-sample confidence intervals for $p_1 - p_2$ and p_1/p_2 in 2 × 2 contingency tables. *Journal of the American Statistical Association, 75,* 386-394.

SAVAGE, L. J. (1962). *The foundations of statistical inference.* London: Methuen.

SCHEFFÉ, H. (1959). *The analysis of variance.* New York: Wiley.

SCHMIDT, F. L. (1996). Statistical significance testing and cumulative knowledge in psychology: Implications for training of researchers. *Psychological Methods, 1,* 115-129.

SMITHSON, M. (2000). *Statistics with confidence.* London: Sage.

SMITHSON, M. (2001). Correct confidence intervals for various regression effect sizes and parameters: The importance of noncentral distributions in computing intervals. *Educational and Psychological Measurement, 61,* 605-632.

SMITHSON, M. (2002). *Confidence intervals for heterogeneity statistics in meta-analysis.* Unpublished manuscript, The Australian National University, Canberra.

STEIGER, J. H. (1990). Structural model evaluation and modification: An interval estimation approach. *Multivariate Behavioral Research, 25,* 173-180.

STEIGER, J. H., & FOULADI, R. T. (1992). R2: A computer program for interval estimation, power calculation, and hypothesis testing for the squared multiple correlation. *Behavior Research Methods, Instruments, and Computers, 4,* 581-582.

STEIGER, J. H., & FOULADI, R. T. (1997). Noncentrality interval estimation and the evaluation of statistical models. In L. Harlow, S. Mulaik, & J. H. Steiger (Eds.), *What if there were no significance tests?* (pp. 222-257). Hillsdale, NJ: Erlbaum.

STEIGER, J. H., & LIND, J. C. (1980, May). *Statistically based tests for the number of common factors.* Paper presented at the annual meeting of the Psychometric Society, Iowa City, IA.

STEIGER, J. H., SHAPIRO, A., & BROWNE, M. W. (1985). On the multivariate asymptotic distribution of sequential chi-square based statistics. *Psychometrika, 50,* 253-263.

TABACHNICK, B., & FIDELL, L. (1996). *Using multivariate statistics* (3rd ed.). New York: Harper & Row.

TUKEY, J. W. (1991). The philosophy of multiple comparisons. *Statistical Science, 6,* 100-116.

VANICHSENI, S., WONGSUWAN, B., Staff of the BMA Narcotics Clinic No. 6, CHOOPANYA, K., & WONGPANICH, K. (1991). A controlled trial of methadone maintenance in a population of intravenous drug users in Bangkok: Implications for prevention of HIV. *International Journal of the Addictions, 26,* 1313-1320.

WALD, A. (1940). A note on the analysis of variance with unequal class frequencies. *Annals of Mathematical Statistics, 11,* 96-100.

WALLEY, P. (1991). *Statistical reasoning with imprecise probabilities.* London: Chapman and Hall.

WILKINSON, L., & APA Task Force on Statistical Inference. (1999). Statistical methods in psychology journals: Guidelines and explanations. *American Psychologist, 54,* 594-604. [Reprint available via APA Home Page: www.apa.org/journals/amp/amp548594.html]

WINDSCHITL, P. D., & WELLS, G. L. (1998). The alternative-outcomes effect. *Journal of Personality and Social Psychology, 75,* 1-13.

YANCOVITZ, S. R., DES JARLAIS, D. C., PEYSER, N. P., DREW, E., FRIEDMANN, P., TRIGG, H. L., & ROBINSON, J. W. (1991). A randomized trial of an interim methadone maintenance clinic. *American Journal of Public Health, 81,* 1185-1191.

ABOUT THE AUTHOR

Michael Smithson is a Reader in the School of Psychology at The Australian National University in Canberra and received his PhD from the University of Oregon. He is the author of *Statistics With Confidence* (Sage, 2000), *Ignorance and Uncertainty* (1989), and *Fuzzy Set Analysis for the Behavioral and Social Sciences* (1987); and coeditor of *Resolving Social Dilemmas: Dynamic, Structural, and Intergroup Aspects* (1999). His primary research interests are in judgment and decision making under uncertainty, social dilemmas, applications of fuzzy set theory to the human sciences, and statistical methods for the human sciences.

Printed in the United States
By Bookmasters